LITTLE VAST ROOMS OF UNDOING

LITTLE VAST ROOMS OF UNDOING

Exploring Identity and Embodiment through Public Toilet Spaces

Dara Blumenthal

ROWMAN & LITTLEFIELD
INTERNATIONAL

London • New York

Published by Rowman & Littlefield International, Ltd.
16 Carlisle Street, London, W1D 3BT
www.rowmaninternational.com

Rowman & Littlefield International, Ltd. is an affiliate of Rowman &
Littlefield
4501 Forbes Boulevard, Suite 200, Lanham, Maryland 20706, USA
With additional offices in Boulder, New York, Toronto (Canada), and Ply-
mouth (UK)
www.rowman.com

British Library Cataloguing in Publication Information Available
A catalogue record for this book is available from the British Library

ISBN: HB 978-1-78348-034-0
ISBN: PB 978-1-78348-035-7
ISBN: eB 978-1-78348-036-4

Library of Congress Cataloging-in-Publication Data

Blumethal, Dara.
Little vast rooms of undoing : exploring identity and embodiment through public toilet spaces /
Dara Blumethal.
pages cm.
Includes bibliographical references and index.
ISBN 978-1-78348-034-0 (cloth : alk. paper)—ISBN 978-1-78348-035-7 (pbk. : alk. paper)—ISBN
978-1-78348-036-4 (electronic)
1. Public toilets—Social aspects. 2. Public toilets—Sex differences. 3. Gender identity. I. Title.
GT476.B58 2014
392.3'6—dc23
2014016000

This book is dedicated to its readers,
through whom new possibilities of being, knowing and
living can be enlivened.

CONTENTS

ACKNOWLEDGEMENTS

This book would not have been possible without all of the amazing encouragement from my chosen families in Canterbury and New York City (and a few outliers in other parts of the world), and my blood family, especially my parents, Ami and Gary, and my brother Asher who are always proud and supportive of my pursuits. A special thanks to Declan, Darren, and Ashley for their continued love, intellectual engagement, and the various senses of home they have provided me throughout the years.

The research and theoretical development in this book were possible thanks to the remarkable support from Chris Shilling throughout the entire process and Dave Boothroyd, who helped me immensely in the final stages of my PhD. I am also grateful to the University of Kent for funding this research and to everyone who participated in it.

BEING (BEYOND) ONESELF

This book is an exploration into the relationship between self-identity and the body as experienced through the ongoing, mundane processes of daily life. In order to explore this relationship I have decided to ground my research in an empirical study of people's experiences of their self-body in public toilet spaces. While self-identity and the body are often understood as having both personal and social aspects, they are most typically theorised as discrete parts of one individual. Public toilet spaces in the contemporary West are thought of in a similarly individualistic fashion. They are understood as both public and private spaces, serving both social and biological needs, yet excretion is an activity normally understood as purely individual, involving just one part of the subject. Once physically and emotionally capable, human beings are expected to carry out their excretory practices as singular beings without further social interference. Interestingly, while these practices are considered wholly private, they are not usually integrated into one's sense of self. Experiences of toileting, while personal, are not thought to give any indication of one's personality or identity. Instead such habitual practices of bodily management serve as a sort of supportive function, a basis for the self which is built onto or out of the highly managed body. Research into public toilet spaces, then, represents an opportunity to highlight the personal and social aspects of the relationship of self-identity and the body that are generally understood as 'second nature' and offer a way into how such supportive functions both maintain and disrupt the self-body relationship in daily life. In what follows, I explicate my exploration into this relationship through three interrelated sections: public toilet spaces as privileged spaces, my theoretical approach to these spaces, and a chapter summary.

WHERE'S THE LOO?

It is important to characterise why public toilet spaces are a privileged example for exploring the relationship introduced above. Put most simply, these spaces are built for bodies. They are associated with both a universal, natural, biological need for humans, i.e., excretion, and society-specific norms, rules, and codes. Public toilets, for the purposes of my study, are those away-from-home, sex-segregated spaces that allow for the urination and/or defecation of at least two persons at any given time. In this definition, not only is the space 'itself' considered 'public', insofar as it is not in someone's 'private' home, but the experience of using the space is also one of publicness, of varying degrees of potential intimate co-presence. The private/public distinction here is not based on the economic delineation of private and public sectors of funding, though there may be some overlap, but is rather directed at the experiential, social conception, and use of the space. Thus public toilets, for the purpose of my project, may be owned and operated by a private institution—like the one that first intrigued me while I was at New York University—or by a business where people go to work every day, in addition to those that are 'publicly' funded.[1]

We know that these spaces are important for living daily life, as social beings, from the likes of sociologists such as Norbert Elias (2000), Erving Goffman (1971, 1977, 1990 [1959]), Sheila Cavanagh (2010), and Harvey Molotch (1988, 2010); urban planners, geographers, and architects such as Alexander Kira (1976), Clara Greed (1995, 2003), Barbara Penner (2009), and Kath Browne (2004, 2006); as well as sanitation and queer rights activists, journalists, filmmakers and public interest lawyers (see e.g., George 2008 and the Sylvia Rivera Law Project at www.srlp.org). We also know that public toilets are important for living daily social life from personal experience. In English-American contexts, with life oriented around waged labour, people are increasingly living in urban areas, and an overwhelming majority leave their home for work, school, household errands, and/or shopping every day. These circumstances mean that the majority of people will by necessity have to use toilets other than those located in their own homes. Indeed, we can safely assume that most people in England and North America have used a public toilet at least once in their life, and a large majority use them every day. While this may be the case, and while we may recognise public toilets as necessary for contemporary social life, little attention has been given to these spaces as places where power operates on, in, and through people—that is, how they not only serve an important function of social life, but how they help enable the reproduction of a normative social experience. So while public toilet spaces may not be central to how we conceive of our daily lives or ourselves—due to the taboo nature of what we do in them, we tend not to give them much experiential currency—they are undeniably important spaces for contemporary life, and it is my suggestion that when we study the experience of the space, rather than the use of the space or the space 'itself', we can garner insights into how we construct embodiment, i.e., the fundamental way we are in and of the world.

Public toilets are part of public, social life, but they are designed both through architecture and social rituals to be used by an individual in undertaking private acts of excretion. These experiences, according to the typical identity structure in English-American contexts, can be described as a juggling act involving one's 'private' self and one's 'public' self. Negotiating the space between public and private usually comes with a toll of personal (1) fear, (2) anxiety, (3) shame, and/or (4) embarrassment directed at one's bodily self. As Goffman explains, such emotions can occur

> whenever an individual is felt to have projected incompatible definitions of himself. . . . These projections do not occur at random or for psychological reasons but at certain places in a social establishment where incompatible principles of social organization prevail.[2]

These four oppressive and repressive emotions are common to public toilet use and are vital elements in this study. Indeed, throughout this book, I am interested as much in *how* individuals *experience* what they do in public toilet spaces as in what people *do* in public toilet spaces. This experiencing of public toilets is never merely an experience of either oneself or the space (i.e., my experience of myself versus my experience of the space), but rather is an amalgamation of one's body in space.

This experience of public toilets begins, I suggest, by identifying oneself with one('s) sex because the spaces are segregated by sex; a factor that explains why this topic has been explored by those invested in identity politics whose projects are typically focused on recognition, representation, and rights of non-heteronormative 'identity categories'. For example, Shelia Cavanagh (2010), in *Queering Bathrooms*, aims to show how public toilets are experienced as threatening to non-normative folk (those who identify as queer, gender nonconforming, trans, etc.) and, oddly, in my opinion, advocates a redesign of the spaces that allows for greater visibility of those individuals who find them threatening. Cavanagh (2010) aims to change people's experiences by changing the space. While I too am interested in how people experience public toilets, it is not my aim to suggest how we should redesign them. Fur-

xiv   **BEING (BEYOND) ONESELF**

thermore, rather than a study steeped in identity politics and oriented towards representation, a politic that is predicated on stable categories of male/female, straight/gay, etc., my research is focused upon a theoretical investigation into the materiality of bodies (of how they are experienced in ways that often exceeds and possesses dissonance with identity categories) and what this reveals about embodied identities. The primary way I accomplish this task is from and through the work of Norbert Elias.

Elias' (2000) study of the European 'civilizing process' keenly traces the development of manners and associated ways of being and thinking about the self-body in society. The development of the European habitus traced through his study, while not specifically body oriented, necessarily implicates the self-body relationship because it socially dictates how an individual should use and be a body in an era of ever-increasing 'publicness' and corresponding 'privateness'. Indeed, this was an era when those two interrelated concepts were first developed and applied to personal habits through social manners and corresponding moral attitudes which were part and parcel of the development of 'public self' and 'private self'. Several of which were explicitly focused on excretory behaviours. It is my suggestion that many of the original social and moral attitudes and related practices, first surrounding excretion and later toileting, continue to shape how we live our daily lives and conceive of ourselves. That is to say that public toilet spaces offer a concentrated and often exaggerated (because taboo) assessment of the self-body relationship that persists today. This is something I explore at length in chapter 4, but it is necessary to sketch here how important Elias is to my study.

Put simply, I use Elias' work for a historical grounding, for organising frameworks of identity, and as an opportunity for the opening of new ideas. This grounding, organising, and opening is the basis for a new ethics of being that I attempt, principally, through a radical re-reading and re-writing of Elias via a posthumanist-material feminist lens. This approach, which I explain below, enables a furthering of Elias' fleshed-out concepts and unrealised desires primarily through a 'new philosophical framework that . . . entails a rethinking of fundamental concepts . . . including the notions of matter, discourse, causality, agency, power, identity, embodiment, objectivity, space, and time.'[3] Central to this framework is a new ethics of being as I explain below.

We know from Elias' *The Civilizing Process* (2000) that there are clear socio-historical elements to how we continue to conceive of and construct individuality in the contemporary West. We also know from Elias' *The Society of Individuals* (1991) that he was deeply unsatisfied with the fact that individuals and societies were often conceptualised as separate and as ahistorical. Following this sentiment I attempt to show how many of the normative, historical ways of being that we continue to reproduce today and take as 'second nature' rather than learned, are unethical according to the new philosophical framework I employ throughout this book. As I explore throughout this book, the social-moral attitudes that continue to underpin many of our social ways of being, and particularly toileting practices, deny some of the most fundamental aspects of being human, including being bodily, caring for others, and a creativity for new ways of living the daily life. What's more, it is of no small consequence that since the built environment of interest here is sex-segregated, those self-experiences are also specifically sex-segregated and give us an opportunity to disentangle the workings of power in a space common to, yet glossed over in, daily life that helps (re)produce an unethical binary sex-gender along a singular axis of heteronormativity, as I address below.

THEORISING TOILETING

This project is theoretically driven and the approach taken both methodologically and generally throughout this book requires some explication. To begin, there are three terms that I use throughout which require defining. Those are *sex*, *gender*, and *heteronormative*. Typically, sex (male/female) is the biological makeup of one's body, and gender (boy/girl, man/woman) refers to those socio-cultural roles and ways of being that are typically based upon one's sex, i.e., those who are female-bodied are expected to be women and thus 'do the things women do'. This process of gendering onto the sexed body renders the materiality of the body into a passive object and the mind into an active subject. Furthermore, those bodies sexed female (and gendered woman) are understood not as *ontologically* different from those bodies sexed male (and thus gendered man), but as an inferior deviation from the ontological human norm of man. This is a classical causal relationship of the body being reduced to mere biological materiality—a *thing*—onto which gender is erected to create a subject. These dialectical relationships, binary sex and its social elaboration into binary genders via the active mind over the passive body, are constructed according to heteronormativity.

Heteronormativity determines *how* we construct gender. Heteronormal constructions of the body are patriarchal and tend to understand the body as abject—something suspect, threatening, despicable and in need of rational management and control, as I explore at length in chapter 5. Heteronormativity thus refers to *ways of being* sexed-gendered on the basis of the oppositional logics of men and women, where, for example, women are passive and men are active, women are bodily and men are mental, women are irrational and men are rational. Traditionally, men, as supposedly rational beings, have been less susceptible to the abject nature of the body, whereas women have not been as lucky, since their minds are not as rationally developed. While this may seem outdated, this binary approach to knowledge and materiality is still at the core of much of our understanding of the world and is a point I develop in chapters 1 and 2. These historical constructions of heteronormative sex-gender have shaped the 'socially accepted', standard (e.g., moral, 'correct', non-threatening, easily identifiable, 'normal') practices of bodies in daily life as briefly explained above. Thus those who are lesbian, bisexual, gay, and queer can still be described as 'heteronormative' insofar as they have an undeniable hetero-gendering historicity, and many of the practices they engage in every day can be described as heteronormative. For example, the different ways girls and boys learn to use their bodies in childhood, which often serves as the basis for their adult forms of embodiment, are heteronormative (Young 2005). Heteronormativity shapes our gender identity and our initial relationship to our body, and can thus be analysed as a material reality which cannot readily be erased from bodily experience. In English-American contexts heteronormativity is central to the historicity and daily experience of identity and the body, and it is through heteronormative constructions of sex-gender that we not only know which public toilet space to use, but also *how* to use it. Put simply, the way our public toilet spaces are split according to sex-gender; how we use our bodies in them; and how we feel about our bodies while using them have all been shaped according to (social-moral) heteronormative ways of being. In other words, toileting practices, which are a set of 'supportive' practices that are considered personal but not related to one's identity, are heteronormative. Therefore, throughout this book, I attempt to show how sex, gender, and heteronormativity inform our materiality in meaning and experience, though often are not recognised as such.

The more general approach taken to and throughout this book which enables the examination of the heteronormative ways of being is *diffraction*. I borrow the term from Donna Haraway (1992), whose concern with 'the way reflexivity has played itself out as a methodology, especially as it has been taken up and discussed by mainstream scholars' prompted her to posit a new optical metaphor for the construction of knowledge.[4] While I explain this in more detail in chapter 3, it is important to situate this approach here. According to Haraway and Barad, 'a diffractive methodology is a critical practice for making a difference in the world. It is a commitment to understanding which differences matter, how they matter, and for whom. It is a critical practice of engagement, not a distance-learning practice of reflecting from afar.'[5] It follows then that diffraction happens at many levels of experience because it allows one to actively recognise the possibilities in and of being through an awareness of material, sensory, and emotive experiences that are normally reduced to 'human nature', minimised through social propriety, or ignored as an unimportant anomaly. This is because, at the core of a diffractive approach is the understanding that matter is always already active, open, and ongoing, while a reflexive approach presumes matter is stable and passive. This is apparent, for example, in how reflexivity is attuned to reflection, mirroring, and sameness, while diffraction is attuned to differential ways of being, living, and feeling. This seemingly simple dissimilarity between a reflective versus a diffractive approach to materiality has major consequences for how we construct and understand knowledge and experience. Accordingly, I outline them below through a sustained meditation on the theory which underpins each of them. I deal first, by way of a brief detour through Deleuze and Guattari's (1994) concepts of chaos and difference, with reflection, and second, with diffraction; thirdly, I give three primary examples of the diffractive approach working throughout this book.

As Elizabeth Grosz interpreting Deleuze and Guattari (1994) explains, 'Chaos is not the absence of order but rather the fullness or plethora that, depending on its uneven speed, force, and intensity, is the condition both for any model or activity and for the undoing and transformation of such models or activities.'[6] Chaos is an understanding of the universe that highlights the innumerable possibilities of being which are available before being named, *identified*. It is only from chaos that we can attempt to create order through models. Models of representation attempt to contain chaos through structures and systems, for example, heteronormative ways of being. Reflection and reflexivity work in accordance with a representational system of knowledge, which only logically works through the essentialising of matter. As theoretical (quantum) physicist cum feminist philosopher Karen Barad explains, 'Representationalism takes the notion of separation as foundational. It separates the world into the ontologically distinct domains of words and things, leaving itself with their linkage such that knowledge is possible.'[7] Reflection and methods of reflexivity rely on this fundamental representationalist separation. For example, not only are we separated and categorised along social understandings (models) of stable sameness of sex-gender (men are one way and women are the opposite [and often inferior] way), this separation requires that things, objects, and subjects are contained within the boundaries of their *own* matter. For example, bodies have boundaries that are sealed. That is to say, at the core of representationalist thinking is an overt wariness of matter. As Barad posits:

> Is not, after all, the common-sense view of representationalism—the belief that representations serve a mediating function between knower and known—that displays a deep mistrust of matter, holding it off at a distance, figuring it as passive, immutable, and mute, in need of that mark of an external force like culture or history to complete it? Indeed, that representationalist belief in the power of words to mirror preexisting phenomena is the metaphysical substrate that supports social constructivist, as well as traditional realist beliefs, perpetuating the endless recycling of untenable options.[8]

In order to work, representationalism employs traditional 'violent forces of mastery, containment, and control posed by masculinist sciences, technologies, and economies' to create an essential separation between ontology and epistemology, between matter and knowledge.[9] While much of this construction of knowledge has been directed at locating the 'true' or 'inner order' of things, that is just 'one mode of addressing chaos, one way of living with it'.[10] As

> Deleuze and Guattari have postulated, beyond the postmodern obsession with representation and discourse, with forms of order and organization, that is, with systems and structures, that philosophy develops nothing but *concepts* to deal with, to approach, to touch upon, to harness, and live with chaos, to take a measured fragment of chaos and bound it in the form of a concept.[11]

Therefore that separation and stability that are necessary for reflection inherently deny the possibilities inherent to chaos.[12] As Grosz explains, chaos

> abounds everywhere *but* in and through the sign. It lives in and as events—the event of subjectivity, the event as political movement, the event as open-ended emergence. The sign and signification, more generally, are the means by which difference is dissipated and rendered tame. Difference is the generative force of the universe itself, the impersonal, inhuman destiny and milieu of the human, that from which life, including the human, comes and that to which life in all its becomings directs itself.[13]

Following this, diffraction is a non-representational approach to chaos that, rather than attempting to locate systems of sameness via reflection based upon representationalist models that do not trust matter, begins by assuming matter is not only trustworthy, but part and parcel of all potential systems of binding chaos, including words and discourse. That is to say, rather than the instability of matter as a source of disturbance that *necessitates* binding, diffraction begins by assuming that the instability of matter is the inherent source of all being, knowing, and understanding in the world. Instability or chaos enables the difference from which any sort of modelling could develop. So, while reflection attempts to create orderly systems of sameness, diffraction attempts to enliven possibilities of generative difference.

   This understanding is important for the study at hand as sameness/difference are central to how we construct identity. According to Deleuze (1994) in *Difference and Repetition*, the concept of difference, in the history of Western thought, has been treated in two troubling ways. It has been evaded, aligned, and repressed, while *at the same time* it has been fundamentally tied to identity, resemblance, and opposition as that which is unconstrained, impossible, and monstrous.[14] While the chaotic possibility of difference is unethically reduced to identity politics through representation and discourse, diffraction situates difference as the ongoing, active *materiality* of knowing and being. The diffractive understanding of difference is not based in orchestrating separation, but is instead invested in *perceiving continuity*. For example, in developing her diffractive approach, Barad asks, 'What if we were to recognise that differentiating is a material act that is not about radical separation, but on the contrary, about making connections and commitments?'[15] Similarly, Elizabeth Grosz explains the relation of difference to identity and identity politics in a diffractive way. She says: 'Difference is the acknowledgement that there are incomplete forces at work within all entities and events that can never be definitely identified, certainly not in advance, nor be made the centre of any political struggle because they are inherently open-ended and incapable of specification in advance.'[16] When considering various identity categories, difference can be used in a representationalist way for the purpose of identity politics, but this is a practice of difference in the name of separation and, as Grosz explains, is not a difference that acknowledges the ongoing nature of being; it is not diffractive. This approach to identity is one of the four philosophical techniques identified by Deleuze (1994) that accomplish the reduction of difference into representation.[17] These four approaches to understanding the world are the

> primary means by which difference is converted, transformed from an active principal to a passive residue. Difference is diverted through identity, analogy, opposition, and resemblance insofar as these are the means by which determination is attributed to the

undetermined, in other words, insofar as difference is subjected to representation.[18]

Thus diffraction is an approach that, instead of reducing difference to a 'passive residue', recognises the ongoing, that is, *active*, differential nature in all phenomena.

Difference according to diffraction is the materialisation of 'the world in its open-ended becoming.'[19] When we understand matter as active, unfinished, and ongoing we can begin to recognise the *entangled* nature of matter and knowledge—we can become aware 'of the apparatuses of production' and enable 'genealogical analyses of how boundaries are produced rather than presuming sets of well-worn binaries in advance';[20] resulting in a situation which exposes a new range of possibilities for living, being, and becoming differently. Thus diffraction as my approach is more than merely a method of research and writing, as through it I am attempting to contribute to the development of a new non-representational knowledge, a knowledge which is non-essentialising of matter. Again, as Barad says,

> Making knowledge is not simply about making facts but about making worlds, or rather, it is about making specific worldly configurations—not in the sense of making them up ex nihilo, or out of language, beliefs, or ideas, but in the sense of materially engaging as part of the world in giving it specific material form.[21]

Language as something outside or wholly separate from matter is central to representational and discursive methods and therefore a reworking of language for this alternative approach to materiality is of great importance. This re-working enables critical engagement with methods of reflection that help produce the boundaries that bind chaos and reduce materiality to passivity. When understood in this fashion, diffraction becomes a method capable of analysing the more simplistic, reductionistic methods of reflexivity. Therefore there are three diffractive methods articulated throughout this book, which I use to draw out differential ways of being. They are: (1) a system of 'technical' expression that attempts to bridge matter and language, (2) the use of 'imaginative' expression that attempts to draw out the material nature of reading, and (3) the overall structure of the book as an example of an entangled phenomenon. I will deal with these in turn. First, the technical terminology used throughout this book attempts to better get at the nature of being in its active, open-ended materialisation. This includes re-appropriated words, neologisms, and expressive combinations such as entanglement (opposed to 'connection' which presumes separation), intra-action (opposed to interaction which presumes separation), and material-discourse (opposed to material/discourse which presumes separation and opposition). For example, and to flesh this out a bit more, my use of intra-action (borrowed from Barad [2007]) serves to highlight the ever-changing nature of 'subjects' and 'objects' of matter. She explains,

> The neologism 'intra-action' signifies the mutual constitution of entangled agencies. That is, in contrast to the usual 'interaction', which assumes that there are separate individual agencies that precede their interaction, the notion of intra-action recognizes that distinct agencies do not precede, but rather emerge through, their intra-action. It is important to note that the 'distinct' agencies are only distinct in a relational, not an absolute, sense, that is, agencies are only distinct in relation to their mutual entanglement; they don't exist as individual elements.[22]

This signals a move beyond 'the assumed inherent or Cartesian subject-object distinction' which believes 'that independently determinate entities precede some causal interaction'.[23] That is to say, rather than individuals with stable identities coming into a space and interacting (or not) with other stable individuals—an understanding that would enable the reflection of norms—'intra-action' considers how (all) elements in an 'interaction' are co-occurring rather than socially prescribed or 'naturally' destined. This enables one to access how norms are actively engaged and reproduced and, in doing so, how a range of possibilities are systematically ignored. For example, let's say a man enters a public toilet space and walks up to a line of urinals, some of which are in use by other men; he doesn't look at the other men or interact with them in any way. Through the lens of interaction, the man in our example engages in non-interaction or what Goffman (1966) may call 'civil inattention'. That is, as Moore and Breeze explain,

> namely a studied disengagement from the space and those within it. We avoid eye contact, act as if we cannot hear or see others and cast our gaze downwards to focus on our own specific path through the space. The benefit of civil inattention in such circumstances is twofold: it is normative and an unmistakeable bodily idiom, to borrow again from Goffman. The rule of civil inattention grants a very narrow range of acceptable behaviour; breaches in the code are obvious and read as a straightforward sign of danger.[24]

The man in this example engages in 'civil inattention' *instead of* interaction. Yet civil inattention is a practice that seeks to *reflect* social norms and thus restricts how one can act, which, when we understand the same example through the lens of intra-action, is important. Through an intra-active approach we can highlight how even 'non-interaction' helps the man in our example (re)produce his *sensation* of normative, individual, masculine identity. That is, it is precisely the non-interaction, the civil inattention that gives relief to the man's 'independent' and 'stable' identity as a man. Intra-action considers how the materialisation of one's 'choice' of behaviour (agency), based upon the entangled nature of a social situation (including the presence of human and non-human elements), is also the materialisation of one's sense of stability as an individual who can express that choice.[25] Therefore, intra-action is a diffractive approach that opens possibilities, by identifying how they are regularly shut down through social rules and codes of interaction.

Second, the use of imaginative or creative expression throughout this book is a diffractive method. More specifically, the use of art, in this case poetry, narrative, and experimental writing, aids in and enables analysis. As Grosz explains, 'Art is that which brings sensations into being when before it there are only subjects, objects, and the relations of immersion that bind the one to the other. Art allows the difference, the incommensurability of subject and object to be celebrated, opened up, elaborated.'[26] The power of art is something many people can easily recognise as already a part of their own experience and thus is important here as a proxy, parallel, or way into becoming aware of the sort of being or embodiment I suggest is possible in daily life more generally. That is, recognising the threshold experiences we already engage in (e.g., through art) may help us to recognise those thresholds we are merely passing over or covering up with socially instituted habit, as I explore in this book. By 'thresholds' I mean the opportunities for feeling, for sensation, for embodied experiences which are not prescribed by the social, which are not pre-digested by representation, and which cannot be *contained* to identity. Thresholds open us to the moments of being which bring us beyond ourselves. Put another way, threshold experiences are de-territorializing.[27] They momentarily interrupt the territorialization of our bodies and of chaos which we attempt to maintain through (practices of) identity. By engaging with the arts we can feel before and beyond ourselves.[28] These forms are also the framing of chaos, like science and philosophy, but instead of slowing down chaos to measure it or binding chaos in concepts to try to create consistency, the arts frame chaos not to control it but to *enable* it.[29] As Grosz interpreting Deleuze explains, the arts 'produce and generate intensity, that which directly impacts the nervous system and intensifies sensation. Art is the art of affect more than representation.'[30] Thus it is through our framing, or territorialization, of our bodies that we are able to experience the de-territorializing effects of the arts, but we have to let those effects in, we have to learn to listen, to be aware, to feel. As I explore throughout this book, I fear we are becoming too rigidly territorialized or too fragmented in contemporary daily life to experience the de-territorializing effects. For this reason, poetry intermingles with the empirical chapters of this book, following the progressively non-normative unfolding of the experiences within the chapters—and indeed the title of this text is also a line of poetry from Marianne Boruch's (2011) *The Book of Hours*. The poetry is here to help spur becomings, to help you feel more, to offer a space for becoming-other and to help recognise how that may feel. The two poems are by contemporary American poet Peter Gizzi and have been selected precisely for their themes of de-territorialization and emergence. While poetry itself, as a form of art, offers a threshold to those who resonate with it, additionally the themes of these poems are also concerned with becoming. The poems explore possibility and differential ways of being at a level commensurate with those experiences analysed in the chapters and expressed through my interview data. In addition to the Gizzi poems, I draw on Kristin Ross's text on French poet Arthur Rimbaud, *The Emergence of Social Space: Rimbaud and the Paris Commune*, and consider both her work and Rimbaud's animating forces in my writing of this book. The poetry, along with text from original performance and personal narrative, as I introduce in more detail below, offers another, imaginative or expressive way into the experiences that are described in the chapters and, accordingly, the chapters enable new insights in experiencing the poems. In this way the art and the analysis are entangled.

The possibility stemming from this entanglement is made more apparent in the epilogue where I've written an interview in a literary fashion. The narrative of the epilogue is directly pulled from one interview, but is presented unlike any of the other data in the book. Here instead of breaking my interviewee's experience apart, I keep his story intact to more fully draw out his experiences of difference, possibility, and becoming-other. Taken together, this 'imaginative' expression in the form of poetry and experimental writing is in line with sociologist Les Back, in *The Art of Listening*, where he declares, 'we have to aspire to make sociology more literary.'[31] My attempt here is to enrich not just the sociological findings through poetry, and the poetry through social experiences, but to enrich the process of reading in its materiality. That is to say, it is not that reading and practices of knowledge merely 'have material consequences but that *practices of knowing are specific material engagements that participate in (re)configuring the world.*'[32] So here poetry is a different material engagement and one that may *make a difference*, that may draw out unexpected feelings and sensations which are elusive and difficult to verbalise, and which may make the sensory-embodied exploration in the following pages more readily material. This is vital, for as Law and Urry explain:

> If social science is to interfere in the realities of [the] world, to make a difference, to engage in an ontological politics, and to shape new realities, then it needs tools for understanding and practising the complex and elusive. This will be uncomfortable. Novelty is always uncomfortable. We need to alter academic habits and develop sensibilities appropriate to a methodological decentring. [33]

So while the poetry may be unusual or surprising, that *in itself* does not detract from the possibilities it may enliven; and this book is focused precisely on extracting possibility from habit.

Finally, the third method of diffraction includes those mentioned previously because it considers the form and content of the book as a whole. While the book has distinct parts, it is, as a whole, an example of diffraction. For example, you will notice that the poems interrupt the very structured progression of the book. The chapters and the content of the chapters are very orderly—mirroring the structuring of habits they elucidate—thus the poems are moments outside of and in between the rigid structure. Put simply, the format of the book itself attempts to materialise the experiential concepts of (de-)territorialization and entanglement I write about. The performative nature of the book is an attempt to diffract or materially draw out differential possibilities into and onto the act of reading and understanding. In its materiality, the book is both a framing of chaos and a de-territorialization of it. One I hope you feel as you proceed.

Ultimately, the case study through which I engage these diffractive methods is the body in public toilets. The two most intriguing factors for me about public toilets are firstly, everyone seems to follow the same rules and codes in there, which means at some point we learned them, and secondly, not only do most people follow the same rules and codes, they also experience the same *feelings while following them*. I'm not claiming that I can access people's feelings, but I can *listen*, and what people have expressed has led me to believe that not only did we learn the rules and codes of bodily usage in public toilet spaces at some point in childhood, but that we also *learned to feel the same negative bodily sensations* and associations when engaged in the rules and codes of public toilets spaces. This observation of sameness—which manifests negative, disconnecting emotions—across bodies is an opportunity for differential becoming. Where we have learned, for example, bodily fear, anxiety, shame, and embarrassment, we can learn to *feel* and *be* differently—not by learning *what* to feel about our bodies but by learning *how* to feel, how to be *consciously embodied*. It is my suggestion that we can *become* before and beyond these negative, normalising, disembodying feelings when we recognise them as habits, which can be released.[34] That is, as thresholds for change. Instead of fear, anxiety, shame, and embarrassment we can learn bodily courage, trust, creativity, and calmness, which can bring us beyond ourselves in a self-generating, cohesive—as opposed to fragmentary and disconnecting—way. These thresholds are momentary and fleeting and they are easy to miss, but they are always already there, we need to learn how to engage them. If nothing else, that is my aim in this book. To put it most simply, in my exploration of identity and the body I aim to show that because we have learned to be one way only means we can also learn to be another way; and not just any way, but to be embodied in such a way that opens us to more and deeper forms of being.

CHAPTER SUMMARY

The three immediately following chapters serve as the theoretical basis of this book. These chapters establish the central theme that flows throughout this book: the relationship between self-identity and the body. The conceptualisation of this relationship is of vital importance for understanding individuality, society, and everyday life. In chapters 1, 2, and 3 I introduce multiple theoretical approaches that seek to make sense of how our bodies and identities interrelate. These approaches are steeped in and stem from the Western philosophical tradition, which conceptualises body-identity through a dualistic (e.g., Cartesian) body (flesh)/mind (identity) model. This portrays the body as passive flesh and a shell that contains the active mind. Even approaches that focus explicitly on the self in experience (e.g., phenomenology) have trouble reconciling this dualism; a dualism that disables us from recognising how the body is more than the sum of its parts, that is, an active and indispensable element to what it is to be a human being.

Throughout chapters 1 and 2 I will expose the limitations inherent in this Western philosophical model and in chapter 4 put forth an approach that seeks to view identity not as separate from or merely expressed through the body, but irreducibly and thoroughly embodied. This approach places the body *as alive*, rather than the mind, at the centre of experience, knowledge, and understanding. Working through this alternative approach (in chapter 3), with the emphasis always on, from, and through the body, my interest is not primarily on how one experiences the body in terms of one's identity (e.g., how a heterosexual woman thinks and feels about her body), but rather how one experiences one's social identity from and through a self that is itself *enfleshed* (e.g., A heterosexual woman not only has but also *is* a body that is inextricable to her social existence. Thinking, feeling, and knowing all happen through her body, not merely her mind situated within her skull.). This may seem a subtle distinction but it is an important one: if we first consider that many people in the contemporary West experience their identity or 'self' as something that exists *within* and not *as* or *through* their body, we can begin to grasp just how different these two philosophical, theoretical, and experiential approaches are. Shifting from thinking that we merely have or possess a body, to the realisation that (a sensory-embodied) identity is only possible *because of* active, ongoing enfleshment, points to new possibilities for conceiving of and also actually *living* individuality and society.

In exploring these approaches—several that are associated with the dominant Western philosophical tradition's separation of the body from the mind and equation of the self with the mind, and my own alternative that seeks to locate selfhood as derived through enfleshed subjects—I am particularly interested in locating how public toileting practices problematise dominant models of identity and can help us see the need to reconceptualise embodied identity. The approach I seek to develop is an empirical elaboration of what Stacy Alaimo and Susan Hekman are calling 'the "material turn" in feminist theory, a wave of feminist theory that is taking matter seriously.'[35] In order to show how different this approach is to conventional views of the body in modernity, I begin (in chapter 1) by focusing on Norbert Elias's analysis of the dominant conception of the embodied self within modernity, the closed, monadic *homo clausus*, and use Leder's (1990) phenomenological account to elucidate this monadic experience. While this way of being is theorised and even experienced as 'natural' or given in the contemporary West, it requires a considerable amount of conscious and ingrained bodywork and social management in order to be sustained. I then introduce (in chapter 2) social constructionist and postmodern approaches that focus on the self as developed from *outside* of one's body.[36] These approaches are in line with Elias's *homines aperti* (men opened) model, but are not entirely successful in moving beyond the ontologically assumed dualisms intrinsic to monadism (i.e., a contained self within a passive, bounded body). Lastly, (in chapter 3) I undertake a more sustained focus on the materiality that informs our bodies and identities by analysing material feminist and posthumanist theories[37] that facilitate a new understanding of the embodied self. My aim here is to provide a theoretical basis for questioning the taken-for-grantedness of *homo clausus* identities that avoids the weaknesses of both social constructionist and postmodern approaches to *homines aperti*. Ultimately, via a detour that involves the development of a posthumanist-materialist lens, I seek to extend Elias's *homines aperti* conception of body-identity by pushing it beyond the inner/outer, open/closed dualisms it retains and the material losses it sustains. In doing so, I offer a Latin neologism that is in keeping with both Elias's work and my post-humanist-materialist approach: *corpus infinitum*, meaning boundless, unlimited, or indefinite body. In doing so, it is possible to productively problematise the 'humanistic model of a subject which has *complete control* over access to knowledge of experience' (Shilling 2003, p. 55, my emphasis) and instead to situate embodied experience and knowledge as alive, ongoing and potentially becoming-other.

These three chapters unfold via three overlapping, entangled sections which trace the conception of self-body experience, knowledge, and understanding from individual independent monad (*homo clausus*), to interdependent fragmented monads (*homines aperti*), to unbounded cohesive sensory-embodiment (*corpus infinitum*). Building on this alternative conception of *corpus infinitum*, these theoretical chapters will then provide the foundation for the following empirically oriented discussions by conceptually and experientially situating *why* and *how* public toileting practices are able to reveal the immense amount of work necessary for maintaining the seemingly stable and normative *homo clausus* identity and for demonstrating where postmodern understandings of the fragmented individual fall short. The three approaches (monadic, postmodern, and becoming) established in these chapters will be employed later as an ideal-typical device that enables us to illuminate and elucidate people's daily experiences. This enables access to sensory-embodied knowledge and understanding that, rather than based on a rigid and simplistic identity structure (*homo clausus*) that is subject to fragmentation and over-stimulation (*homines aperti*), is pliable, cohesive, and indefinite (*corpus infinitum*), because it takes the materiality of the body as ontologically primary and always already actively open to change. Thus *corpus infinitum* offers a way into experiences of becoming-other, of opportunities for differential ways of being. This signals a move away from the classical identity structure based on bodily dominance and control as the root necessary for knowledge, to new ways of conceiving of and recognising material living as the inherent source of it.

Following the theoretical section is a short historical elucidation of the development of public toileting spaces. Therefore, in chapter 4 I outline the history of public toilet spaces since the fifteenth century through three spatial-historical milestones. Those milestones include both the social and the architectural developments of public toileting and are elucidated according to their entanglement with the normative identity structure they support. I explicate these milestones through a sustained focus on Elias's (2000) *The Civilizing Process* through which I consider the development of manners and bodily ways of being associated with increasing levels of shame and embarrassment of the body. Central to this development is the progression of privacy from the communal and thus results in the development of public and private as parts of the social as well as parts of the person. In this chapter I show how the development of privacy and the public toilet are implicated in the development of individuality as the standard model of identity construction in the West. Both the spaces of public toileting and the associated attitudes, as I explore in this chapter, have not changed much since they were solidified in the Victorian era and thus are continually reproduced in our daily experiences and expectations of the body and social propriety. Following this chapter is the section of empirical findings as I discuss below.

The three empirical chapters of this book, chapters 5, 6, and 7 correspond with the three theoretical chapters described above insofar as they each explore, through empirical, experiential data, one of the primary themes laid out in the theory chapters. Specifically, chapter 5 is an empirical elucidation of chapter 1, chapter 6 corresponds to chapter 2, and chapter 7 explores the approach taken in chapter 3. In order to explicate the three different approaches to self-body identity through my empirical data I have developed the idea of an 'intra-action order' of public toilets that enables me to analyse what I depict as normative experience while also giving relief to the non-normative and threshold experiences contained in my data. In order words, through the territorialization or structure of public toileting behaviours in chapter 5, I can more readily access both the de-territorializing behaviours as well as the opportunities for becoming-other through a system of 'intra-action' in chapters 6 and 7. I will outline this in more detail below.

Chapter 6 introduces the 'triadic intra-action order' (TIO) of public toilets based on the *homo clausus* experience of the abject body. I use Kristeva's (1982) concept of abjection to elucidate the *homo clausus* approach to the public toileting body as something untrustworthy and in need of control. My intra-action order is developed from and in response to Erving Goffman's (1983) interaction order, through which he analyses the social dynamics of co-presence. The TIO operates according to three entangled rules. They are: minimise your movement, mind your eyes, and manage your boundaries. The TIO refers to a set of practices that are embodied from a young age and through repetition appear natural and universal. They are a conditioning of the body that is experienced as separate from the self. Each rule of the TIO is elucidated through data from my interviews and/or surveys and includes experiences of men, women, genderqueer and trans individuals. The rules of the TIO apply to both men's and women's spaces though often operate in seemingly opposite ways. This contributes to how heteronormative *homo clausus* sex-gender is constructed along a singular axis of ontology with two 'different', indeed opposite, versions—male and female—of the same bodily need. I also explore in this chapter how learned experiences of fear, anxiety, shame and embarrassment perpetuate the TIO and help *homo clausus* maintain their sense of individuality.

Chapter 6 introduces practices of embodiment that correspond to *homines aperti* ways of being. This chapter explores how bodies are interconnected and interrelated in public toilet spaces. The practices focused on in this chapter are thematically related to practices of care. These practices of care include self-care and caring for others and are transgressive insofar as they explicitly challenge one or more rules of the *homo clausus* triadic intra-action order which attempts to stabilise *homo clausus* body-identity through experiences of bodily fear, anxiety, shame, and embarrassment. Practices of 'care in toileting' explored in this chapter work to overtly expose the inherent openness and interconnectedness of bodies by highlighting their vulnerability and in doing so can reveal how the monadic confines of *homo clausus* norms are contingent and frail rather than universal and stable. I also explore how, while fear, anxiety, shame, and embarrassment continue to play a large role in the practices of *homines aperti*, practices of care can help expose the fissures of *homo clausus* individuality, and thus better enable us to access thresholds for greater intervention in sensory-embodied experiences.

Chapter 7 is the final empirical chapter (the epilogue is also empirical, but is presented is a very different way). Building on the developments gleaned in chapter 6, chapter 7 pushes my analysis of the fissures in the *homo clausus* identity structure even further. The stories in this chapter coalesce around themes of play, pleasure, and possibility, and work to expose how one's bodily being in and of the world can shift from rigid habit to open, boundless becoming. These experiences explicate the possibility of self-bodily identity to be described and experienced as *corpus infinitum*. In this chapter I explore the limits of *homo clausus* habit ordered by the triadic intra-action order and show how habits can be dissolved over time, allowing for new ways of being to have an a/effect. This chapter explores how differential ways of being are part and parcel of all practices, and thus expose how the rules of the triadic intra-action order are not immutable, but rather contestable and easily ignored. The practices of play, pleasure, and possibility in this chapter point to the potential to experience a self outside of the *homo clausus/homines aperti* dialectic of understanding and point to the possibilities of being inherent to a material self which is active and always already open to the experiences of becoming-other. Along with interview data, I also use text from a piece of performance art entitled *Tearoom Sympathy* (written and performed by one of my participants) in order to expose the ability to experience a self not restricted by *homo clausus* heteronormative ways of being. The practices presented in this chapter explore how when experiences of fear, anxiety, shame, and embarrassment are actively ignored or productively utilised for the opening, that is, the de-territorializing of one's self-body, the power of those emotions to regulate behaviour is radically diminished. These practices seize the fissures inherent to *homo clausus* (in)stability and expand them through new ways of being. This results in a radical politic of bodily being, beyond passive habit, anxious abjection, and rational control. That is, a new ethics of being.

Overall, throughout this book I explore how people experience their bodies in public toilet spaces and how that impacts their sense of self. I study how when ideas of the self are too rigid, when the body is too territorialized, the powerful effects of de-territorialization, of active materiality can barely be felt. Likewise, I investigate how when one is seemingly too fragmented, too accustomed to change, that is, too de-territorialized, there is little room for growth and change through a sensory-embodiment that is not considered active. In order to problematise this dialectical approach to identity as open or closed, inner or outer, I develop an approach which seeks balance between the two. Thus, this book isn't about a new structure of identity, but about the ability to recognise the myriad potentials available to us as bodily beings. It is my hypobook that when we learn to trust matter, to recognise its inherent and active part in all of our entangled phenomena (e.g., thinking, knowing, understanding, experiencing), we can allow ourselves to be open to becoming beyond those habits and ways of life that prop up social structures based on passive, stable matter. Thus we open ourselves to embodiment, to threshold experiences, to becoming-other. Rather than a collection of stories and anecdotes that tell us something about social life, the following chapters contain experiences of habit, struggle, and becoming-other that tell us something about being human, about having and being a body in society. It is my suggestion then, that public toilet spaces can help us access the workings of power usually ignored in daily life, and thus help us to recognise the need to reconceive of how we construct and understand identity and embodiment in our (re)configuring of the world.

NOTES

1. In both the American and British contexts, the distinction between privately owned and operated versus publicly owned and operated is often unclear. For example, in New York City, there are governmentally cared-for public toilet buildings on the premises of city parks, which are free to use and are funded through public-private partnerships (i.e., government funding and private funding), and in England there are department stores, which have pay entry public toilets—these can be considered public toilets on private property that fetch a public fee for use.

2. Erving Goffman, 'Embarrassment and Social Organization', *American Journal of Sociology* 62, no. 3 (1956): 264.

3. Karen Barad, *Meeting the Universe Halfway: Quantum Physics and the Entanglement of Matter and Meaning* (Durham, NC: Duke University Press Books, 2007), 26.

4. Barad, *Meeting the Universe*, 29.

5. Barad, *Meeting the Universe*, 90.

6. Elizabeth Grosz, *Chaos, Territory, Art: Deleuze and the Framing of the Earth* (New York: Columbia University Press, 2008), 26–27.

7. Barad, *Meeting the Universe*, 137.

8. Barad, *Meeting the Universe*, 133.

9. Elizabeth Grosz, *Becoming Undone: Darwinian Reflections on Life, Politics, and Art* (Durham, NC: Duke University Press, 2011), 150. When understood in this way any knowledge based on representation is indebted to patriarchal ways of being.

10. Grosz, *Chaos, Territory, Art*, 27.

11. Grosz, *Chaos, Territory, Art*, 27, original emphasis.

12. As Grosz (*Chaos, Territory, Art*, 27) explains, 'This concept of chaos is also known or invoked through the concepts of: the outside, the real, the virtual, the world, materiality, nature, totality, the cosmos, each of which is a narrowing and specification of chaos from a particular point of view. Chaos cannot be identified with any one of these terms, but it the very condition under such terms are capable of being confused, the point of their overlap and intensification.' To this list I would add Irigaray's 'sexual difference', my use of 'embodiment', French poet Arthur Rimbaud's 'laziness' and 'Bourdonnement', and philosopher Friedrich Nietzsche's 'swarm'.

13. Grosz, *Becoming Undone*, 94, original emphasis.

14. Grosz, *Becoming Undone*, 92.

15. Karen Barad, 'Quantum Entanglements and Hauntological Relations of Inheritance: Dis/Continuities, SpaceTime Enfoldings, and Justice-to-Come', *Derrida Today*, 3(2), pp. 226.

16. Grosz, *Becoming Undone*, 94.

17. Grosz, *Becoming Undone*, 93.

18. Grosz, *Becoming Undone*, 93.

19. Karen Barad, 'Posthumanist Performativity', in *Material Feminisms*, ed. Stacey Alaimo and Susan Hekman (Bloomington: Indiana University Press, 2008), 139.

20. Barad, *Meeting the Universe*, 30.

21. Barad, *Meeting the Universe*, 91.

22. Barad, *Meeting the Universe*, 33, original emphasis.

23. Barad, *Meeting the Universe*, 130–31.

24. Sarah Moore and Simon Breeze, 'Spaces of Male Fear: The Sexual Politics of Being Watched', *British Journal of Criminology*, advance access published August 9, 2012, doi:10.1093/bjc/azs033: 6.

25. This is not to say that matter is materialised ex nihilo through intra-action, but rather that matter is always already an ongoing and active process of being and becoming beyond identity; the sensations with which we *identify* happen in that process.

26. Grosz, *Chaos, Territory, Art*, 78.

27. Gilles Deleuze and Félix Guattari, *Anti-Oedipus*, trans. Robert Hurley, Mark Seem, and Helen R. Lane (New York: Continuum, 1972).

28. Following Grosz (*Chaos, Territory, Art*, 3), here I too am interested in 'all forms of creativity or production that generate intensity, sensation, or affect: music, painting, sculpture, literature, architecture, design, landscape, dance, performance, and so on.'

29. Grosz, *Chaos, Territory, Art*, 27.

30. Grosz, *Chaos, Territory, Art*, 3.

31. Les Back, *The Art of Listening* (New York: Berg Publishing, 2007), 164.

32. Barad, *Meeting the Universe*, 91, original emphasis.

33. John Law and John Urry, 'Enacting the Social', published by the Department of Sociology and the Centre for Science Studies, Lancaster University, Lancaster LA1 4YN, UK, at http://www.comp.lancs.ac.uk/sociology/papers/Law-Urry-Enacting-the-Social.pdf, 11.

34. This phrase connotes an engagement with those aspects of being animal (e.g., sensation) prior to reorganisation through humanistic models of being (e.g., identity) as well as those phenomena that bring us out of our humanistic understandings to new and different post-human experiences.

35. Stacy Alaimo and Susan Hekman, *Material Feminisms* (Bloomington: Indiana University Press, 2008), 6.

36. I use 'postmodern' and 'poststructuralist' to generally refer to theoretical approaches that are primarily linguistic, invested in discourse and seeking to further an understanding of identity as something which is individual in the former and fragmented in the latter. These approaches tend to focus on the way social structures work on and produce individual bodies, making the body a locale for theory and understanding, but not accounting for the body as the generative source of knowledge. Inherent to this is an understanding of humans as singular, individual, bounded entities. For a sophisticated analysis of such approaches that is in keeping with my own approach see Susan Hekman, *The Material of Knowledge: Feminist Disclosures* (Bloomington: Indiana University Press, 2010).

37. It is important to note that many of the material feminist and posthumanist theorists I engage with here are heavily influenced by science studies and a Latourian point of view. Karen Barad, one of the most important feminist theorists contributing to this school of thought, obtained a PhD in particle physics, which she taught for many years before moving into more interdisciplinary, feminist, and philosophical work.

I

The Dis-Embodiment of Identity

I

HOMO CLAUSUS AND THE WESTERN PHILOSOPHICAL TRADITION

Walls, then, are built not for security, but for a *sense* of security. The distinction is important, as those who commission them know very well. What a wall satisfies is not so much a material need as a mental one. Walls protect people not from barbarians, but from anxieties and fears, which can often be more terrible than the worst vandals. In this way, they are built not for those who live outside them, threatening as they may be, but for those who dwell within. In a certain sense, then, what is built is not a wall, but a state of mind.

—Costica Bradantan

This chapter will introduce and set out the main features of Norbert Elias's model of *homo clausus*: the conception of the body-identity relationship that he identifies as dominating the Western philosophical tradition and as reflecting how many people experience and live their bodily being in modernity.[1] *Homo clausus* is the closed, monadic subject who has a high degree of rational, emotional, and physical self-control. As independent individuals, they are separated from others by the borders of their physical selves and are assumed to be autonomously in control of their bodily being. These autonomous individuals seemingly have no bodily or social history. As Elias explains, 'The concept of the individual is one of the most confused concepts not only in sociology but in everyday thought too. As used today this concept conveys the impression that it refers to an adult standing quite alone, dependent on nobody, and who has never even been a child.'[2] In line with this autonomous individual who does not have a childhood (during which social ways of being embodied were naturalised, as I explore below), *homo* *clausus* is assumed to possess a basic, essential identity that exists prior to and remains significantly untouched by social interaction. Instead of being differentiated on the basis of the material or lived differences deriving from such variables as sex, gender, and sexuality, the *homo clausus* model of the embodied self assumes a universally knowable, 'neutral', and stable *subject* that *has* or *possesses* a passive material body. *Homo clausus* subjectivity thus functions through a high degree of bodily disconnection, control, and management. This work enables one to rationally hold together an individual self-identity, which gives the impression of body-identity stability; even though one's physical experience may not be so stable. My suggestion here is that the *homo clausus* subject is encouraged and taught to know himself *rationally* not physically or sensorially, because all relevant experience, knowledge, and understanding (supposedly) happens through his mind. This body-identity is steeped in a *naturalised* experience of the individual self, existing *within* the sealed borders of the passive body and therefore is a dualism that is experienced in daily life as singular and monadic and teaches one to try to ignore, suppress, or 'correct' any physical or emotional differences that do not fit this conception of individuality. *Homo clausus* is not just a collection of social norms, but also, a matter of how the body is socially and individually *experienced* as an entirely separate, self-same, sealed-off entity. *Homo clausus* is thus, most fundamentally, a style of *dis-embodiment*.

INDIVIDUATING BODIES

To arrive at and maintain *homo clausus* identity in daily life involves the condensation and reproduction of several personal and social processes that attempt to stabilise it. As conceived by Chris Shilling, *homo clausus* identity is made possible through the development of three primary bodily characteristics: Socialisation, rationalisation, and individualisation.[3] Firstly, socialisation involves making bodies 'socially acceptable' through the specific embodiment and 'expression of codes of behaviour' and 'the hiding away of natural' and biological functions through personal, spatial, and technological means.[4] Through the distancing and denial of natural and biological realities the body increasingly becomes associated with the social.[5] While Elias initially located this socialisation process on a historic scale, developing over many centuries, as I highlight in chapter 5, it is for contemporary individuals a condensed process, occurring over a matter of months and years in one's childhood. This social control that everyone in modern Western society is expected to go through, is experienced as learning self-control. The most obvious example relevant to this study is the process of toilet training, through which small children learn to discipline and control their natural, biological functions through the space and technology of a 'private' toileting facility. In order to become a recognised member of Western society one must express control over excretory functions. This process, among other things, teaches one that the body is fundamentally biological, is something to control and manage *in order* to have a human identity.

Secondly, *homo clausus* is also an outcome of corporeal rationalisation; a rationalisation that also involves a physical differentiation as the body 'is seen as less of a "whole" and more as a phenomenon whose separate parts are amenable to control.'[6] By learning to give attention to the body in terms of separate parts, one learns to rationally discipline and control oneself in what seems like discrete 'mind over matter' expression. Through rationalisation, the body is conceptually (and thus experientially) broken down and pieced out, allowing it to be described through modern mechanical metaphors which render the body into nothing more than the sum of its parts. This process of rationalisation speaks directly to what I term 'sensorial individuation', i.e., a mode that allows us to conceptually separate and count our sensorium and apply cultural value and meaning to some 'senses' over others. Consider how our rational counting of 'the senses' (that being five) differentiates them, makes them *knowable*, but does not necessarily make them separate in *experience*.[7] In the contemporary West we conceptualise 'normal' able-bodied persons as having the capacities for seeing, hearing, tasting, touching, and smelling as discrete senses, not a cohesive sensorium. Furthermore, we give little attention to proprioceptive and/or kinaesthetic sensoriums and label balance, for example, as something that someone either 'keeps' or 'loses' and the degree of spatial awareness someone may possess (e.g., how clumsy they might be) as part of their 'personality' or 'genetics', erasing the fact that we *learn*, firstly in childhood and as a continual process, how to use our bodies in space and how to move around in the world. Rather than a source of knowledge, expression, or oddly identity, the separated senses and partitioned body become tools for rationality through which passion, pleasure, and violent emotional expressions are mediated and controlled. Developing 'sensorial individuation' and drawing on Asia Friedman's 'selective attention' and 'optical socialisation,' I suggest below that the sense of sight is one of the most intensely rationalised senses and extremely vital for the maintenance of *homo clausus* identity.[8] This process of rationalisation necessitates mental disconnection from, learned ignorance towards, or oversimplification of the physical and sensorial. It is a system of categorisation.

Processes of *consciousness* are often necessarily processes of *sensing* (and are always material processes, as I explore in chapter 3), but sensory-embodiment is ignored or left out of the rational construction of knowledge, understanding, and 'sense making'. That is to say, consciousness is not merely rational, but an entangled embodied process which is crudely condensed into the rational. Consciousness *is* material. Thus 'rational processes' often leave out the embodied aspects and are taken to be separate from or the basis for other 'rational processes'. As Elias explains, 'The idea that what we reifyingly call "consciousness" is multi-layered is the outcome of an attempt to set up a new mental framework within which specific observations can be processed and that can serve as a guide for further observations.'[9] Observations are crucially sense-based happenings. In order to *perceive*, one must engage one's embodied sensorium. It is only through this foundational 'mind-over-matter' split that we can continue to produce dualistic, rationalistic observations emblematic of the Western philosophical tradition. The rationalisation of the body not only constructs the imagined borders of the mind, i.e., within the skull, but also firmly locates thinking, knowing, and understanding within those borders by neglecting to become aware of, or include sensory-embodied experiencing. The *homo clausus* individual knows experience only through the rational mind, not the sensate body. Those rational processes are understood to be something other than material, and since it is the basis of understanding, *homo clausus* is a conditioning of materiality that seeks to produce this same binary amongst all bodies. It is a functional procedure of sameness from which representational difference can be ascertained.

Thirdly, the individualisation of the body completes the monadic *homo clausus* identity by supplying it conceptually and experientially with an 'outer' boundary (i.e., the skin) that separates the 'individual' from other 'individuals' within 'society'. As Elias points out, 'The idea of the "self in a case" is a central theme within modern Western philosophy' and vital for the dualisms raised above.[10] As explained by Mennell, 'Descartes (1596–1659) played a major role in establishing the tradition of philosophy which is preoccupied with a consciousness of one's own consciousness, and striving to understand one's own understanding, as a single adult mind, *inside*, striving to grasp by Reason the problematic world *outside*.'[11] This philosophical basis has seeped into daily life and forms the basis of how many people understand their experience within Western capitalist individualism—that is, through a false awareness of consciousness as separate from the body. That is to say, we are still suffering the effects—both in the production of philosophy, knowledge, science and in mundane daily life—of Cartesian dualism that completely overlooks the process of how one's consciousness of being conscious is only possible through active, ongoing embodiment. Individualisation creates the imagined borders and boundaries, between the 'inside' and 'outside', of one's body, with the outermost layer of skin acting as the sealant. While Descartes takes it as a given that consciousness lives within one's body, it is only possible through an ongoing process of disembodiment. In order to create an internal sense of self, the body has to be hemmed in and made distinct from *other* bodies in the 'external' world. This is necessary to give the impression that one has a stable sense of self, an identity that is separate and distinct from those selves around them. Together, these three interrelated and interdependent processes enable the *homo clausus* self-body-identity.

DE-SEXING THE MONAD

Before continuing my general elucidation of *homo clausus*, it is important to briefly address the relationship of sex and gender to this monadic self-body-identity. In reflecting the dominant Western philosophical tradition's focus on the mind as that which defines us as humans, *homo clausus* is founded upon a heteronormative view of the male, heterosexual body as that which approximates closest to its autonomous ideal. This ideal serves as the ontological basis for all bodies regardless of sexual difference. Historically, this has been extremely problematic for those sexed as female and gendered as women because menstruation, pregnancy, birth, and menopause are fundamental to sexual, that is *ontological*, difference but are not the types of bodily functions that are easily inculcated into the rational *homo clausus* system of identity and knowledge; and let us not forget that feminine sexuality has a long history of being completely ignored, obfuscated or stigmatised through religion, science, and medicine.[12] This is how the self-same 'neutral' (i.e., male) body-identity has been perpetuated for so long. Emily Martin brings these discrepancies to light in her study of American women in which she asks women *not* to talk about 'their families, spouses, and children (when they seem very likely to simply reproduce a version of dominant cultural ideology) but about themselves, through the medium of events which only women experience and which perhaps for that reason are rarely spoken of—menstruation, childbirth, and menopause.'[13] She points out that 'women are not only fragmented into body parts by the practices of scientific medicine, as men are; they are also profoundly alienated from science itself.'[14] Since women have been historically and systematically excluded from the process of knowledge making, their ability to express their 'consciousness of being conscious' (in the Cartesian sense) has been socially limited or entirely unrecognised. Therefore 'the depiction of modern consciousness leads to the conclusion that women's lives are especially degraded, fragmented, and impoverished.'[15] Sexual difference is taken not as the fundamental difference of a different human experience, but as indicators of weakness and inferiority which ultimately mean they are lesser than the ideal man. Even where women attain positions of power within patriarchal hierarchies, their bodies are still considered to 'interrupt' or 'interfere' with their 'mental' capacities. Thus, no matter how strictly those sexed female adopt and adhere to *homo clausus* norms, their bodies will always be transgressive because their physical realities and their *'perceived'* (i.e., presumed) mental capacities are limited because of their biology and thus do not match those of the *homo clausus* ideal—that is, an ideal of sameness across all bodies. It is only through the questioning and exposure of this taken-for-grantedness that new models can be developed. This cause is taken up by queer and feminist theorists, such as Judith Butler (see, e.g., 1988, 1993, 1999 [1990], 2004), who sought to disrupt heteronormative assumptions of sex, gender, and desire, as I will address in chapter 3 when interrogating social constructionist and postmodern approaches.

CONTAINING AND CONTROLLING THE MONAD

As explored above, the development of the *homo clausus* identity structure involves the naturalised splitting of the mental and physical functions through society-specific propriety. We learn how to be and have bodies based on societal systems of rationality and dis-embodiment; this is a point I develop at length in chapter 5. While this may seem like 'human nature' (i.e., innate), it is instead the outcome 'of a long civilizing process in the course of which the wall of forgetfulness separating libidinal drives and "consciousness" or "reflection" has become higher and more impermeable.'[16] This sense of 'human nature' is continually reproduced in contemporary societies because 'the adaptation of young people to their adult functions usually happens in a way which particularly reinforces such tensions and splits within the personality.'[17] Maintaining the *homo clausus* way of being requires a high degree of self-consciousness[18] and self-control (which falsely gives the impression of complete mental understanding of and control over experience while ignoring that those *are* indeed material processes). Elias speaks explicitly about how the battles that used to take place between people, for example, now more frequently take place within individuals; as the supervising elements of one's consciousness struggle to keep drives and desires from 'breaking through' one's 'bounded' flesh or spilling out into the public realm, as that would signify a loss of control and a loss of the borders that contain the individual. This, no doubt, results in a loss of awareness of one's drives and desires and instead becomes a consciously controlled way of being.

Without this work the *homo clausus* subject could not maintain a normative, legible body-identity in society. Through this self-consciousness and self-control the *homo clausus* subject learns how to *have* a body, rather than how to *be* embodied. The body is merely a biological given, a supportive apparatus for the mind (i.e., self) that requires strict physical and emotional management. This management of the fleshy, biological, and emotive body is apparent through not only outward social manifestation, but also bodily patterns on the more subtle, personal level. This is clear if we compare characteristic modes of breathing that separate individuals in the contemporary West, from those Eastern practitioners of yoga or tai chi chu'an analysed by Mauss (1973). Individuals brought up in the domestic cultures of the West tend to breathe shallowly, using only the upper portion of their lungs and restrict their breathing in daily life when engaged in certain activities, whether that be reading e-mail, in conversation, or driving to an appointment. This is also clear in how people deal with stress that cannot be emoted freely, by holding it in their muscles by tensing, e.g., their shoulders, jaws, and/ or stomachs, or constantly shaking a foot. Similarly, *homo clausus* subjects are inculcated to employ certain emotional states to hold together this body-identity.

As Elias's elucidation of *homo clausus* suggests, deeply ingrained experiences of fear, anxiety, shame, and embarrassment (FASE) overlay these bodily processes and disciplines and are vital for maintaining the sense of sameness and stability in this individual body-identity. FASE are deeply connected to the development and continued engagement of self-consciousness and self-control and are four emotions I focus on throughout this book. These states of being are experienced as part of 'human nature' but how, why, when, and where we experience them are taught and learned in society-specific ways. FASE are so powerful for the *homo clausus* identity because they are socially instilled to reinforce the 'self in a case' experience. To paraphrase Elias, these emotions arouse the feeling *within* an individual that one is separate from other people.[19] FASE seemingly deny the connectivity of people and social life and often make us feel like shrinking, hiding, or fleeing. Learning to experience these states as individuals props up and increases the seemingly palpable divide between one's self and others.

FASE are basic and crucial aspects of *homo clausus* identity because individuals are not only encouraged to experience these emotions but to also impose them upon *themselves* during occasions when they fail to live up to the standards of the isolated, ideal monadic subject. This imposition requires both self-consciousness and self-control. Should someone fail to engage in or display such self-centred characteristics, they are thought to lack the level of maturity or rationality of the stable, controlled *homo clausus* subject. They are thought to be abject and *different*. FASE in relation to the *homo clausus body* are of particular interest here. When it comes to these emotions as based upon the fleshy, biological body, it is often difficult to rationally understand the emotions separately. A general sense of fear and anxiety (e.g., unease, worry, concern) regarding bodily transgression work to keep the *homo clausus* subject mentally aware of and in control of bodily borders (to the extent possible) at all times. When control cannot be maintained or must be relinquished (e.g., during excretion, sex, or a heated argument), the subject risks shame and embarrassment. FASE are related to the ability of one to maintain their *homo clausus* identity as stable and socially legible. Fear and anxiety, in relation to the body, are emotions entangled in social connectivity. One may fear or be anxious about social exclusion and ridicule based on embarrassing or shameful instances or possibilities. Shame and embarrassment are some of the first emotions we learn to feel in relation to our bodies; we are taught to control, manage, and discipline our bodies because otherwise they will bring about embarrassment and shame. Thus instead of learning to *feel* our bodies as entangling (with our) minds, we learn to feel these emotions. They act as a self-body buffer; a rational-emotional system used to keep the body merely biological, untrustworthy, and disconnected from the self. What's more, these emotions are often considered juvenile and can make one feel young or small because they are emotions learnt early in life, when one is learning to discipline one's body according to *homo clausus* subjectivity, and thus we associate them with our 'base' bodily needs. They are both inherently degrading and necessary to function socially.

Erving Goffman, who describes how *homo clausus* ways of being operate in social life (as explored in the following chapter), has given thorough attention to the topic of embarrassment. He describes it as

> a possibility in every face-to-face encounter. . . . It occurs whenever an individual is felt to have projected incompatible definitions of himself before those present. These projections do not occur at ran-

dom or for psychological reasons but at certain places in a social
establishment where incompatible principles of social organization
prevail. In the forestalling of conflict between these principles, embarrassment has its social function.[20]

As Goffman describes, embarrassment acts a sort of buffer zone within
an individual. Because certain bodily expressions are not socially acceptable, yet individuals must still be (bodily) in public, embarrassment
helps create distance between these two incompatible features of the
social. Goffman explains how individuals are understood to have one
body and thus engage in one set of performances which expresses one's
self. These characteristics are generally static and innate and do not
allow one to have different personalities, but rather one enduring (i.e.,
highly managed) identity. Furthermore, according to Goffman's inter-
pretation, one is expected to have complete agency over one's bodily
'projections'—that it is one's duty to be in control of the way one's
identity is 'given off' and 'sent out' into the world. He believes that
individuals can and should be able to manage this process and, by using
FASE, help that management process.

    Similarly, Scheff, who posits that experiences of shame exist at a low
grade in all social interaction, understands shame to be *'caused by the
perception of negative evaluation of the self.'*[21] Instead of the belief that,
in modern societies, shame is an emotion that adults rarely experience
(as evidenced, for example, in anthropology and psychoanalytic theory),
Scheff believes, similar to Goffman's convictions regarding embarrassment, that shame is always potentially present in social life.[22] He explains that 'shame is *the* social emotion, arising out of the monitoring of
one's own actions by viewing one's self from the standpoint of others.'[23]
(This is also a prerequisite for *homines aperti* and is explored in detail in
chapter 3.) The experiencing of these emotions is socially instituted and
individually imposed. These emotions help create and maintain naturalised barriers between bodies and reinforce *homo clausus* subjectivity as
separate from and merely expressed through the body. Instead of an
embodied experience of one's self, this embarrassed and shameful subject is created by projecting awareness or consciousness (oneself) outside of and back onto the surface of the body; confining the emotion to
the internal parameters of the enfleshed self. *Homo clausus* subjectivity
is based upon this ability to 'naturally' monitor the borders of one's
body from one's rational awareness and not allow anything to leak out of
the 'sealed' body.

    'Naturalness' is generally associated with comfort and 'unnatural-
ness' may refer to embarrassment, shame, or other type of discomfort.
As Goffman explains:

> In the popular view it is only natural to be at ease during interaction,
> embarrassment being a regrettable deviation from the normal state.
> The individual, in fact, might say he felt 'natural' or 'unnatural' in the
> situation, meaning that he felt comfortable in the interaction or em-
> barrassed in it. He who frequently becomes embarrassed in the pres-
> ence of others is regarded as suffering from a foolish unjustified
> sense of inferiority and in need of therapy.[24]

In Goffman's conception, 'naturalness' is conflated with 'normativity'.
This is similar to the way that heteronormative *homo clausus* identity is
conflated with the biological body. FASE help maintain this sense of
'naturalness' in the reproduction of *homo clausus* body-identity. They
help prevent the loss of monadic boundaries because they act as the
body's defences from the outside-in. Fear, anxiety, shame, and embarrassment, through rational awareness, produce the boundaries and barriers of one's physical self because they require one to experience self-
awareness from an imagined outside (i.e., self-consciousness). The social taboo and stigma associated with these negative feelings also keep
subjects from experiencing other aspects, emotions, sensory processes,
movements—in short, phenomena inherent to embodiment. FASE regulate how the body can be used in everyday life according to naturalised
normativity. While FASE may be experienced as uncomfortable or
even painful, they are not natural, but rather learned barriers imposed
onto the fleshy-self that help activate a very specific sense of inner and
outer worlds. When experienced, these emotions give the impression
that the body is bounded and sealed, while actually revealing the intense mental and emotional work that goes into holding together the
homo clausus body-identity. These negative feelings rationally construct
and distance one's sense of self from one's bounded body in a cycle that
ensures the continuation of FASE. Since this cycle begins at such an
early age (e.g., with toilet training) the borders of the body/self seem
natural and stable, as do the associated embodied emotions. Through
the early embodiment of FASE we learn to stabilise and mentally distance 'our-self' from our (and other) bodies.

LOOKING OUT FROM WITHIN

Bodily fear, anxiety, shame, and embarrassment have a very clear, socially distinct, rational basis that one typically begins to learn in early
childhood and is a process that continues throughout adolescence. It is
often through these negative feelings that people first learn how and
where to conceive of the borders of their bodies and therefore what

they *should* feel responsible for as *their own*. It is an overarching process that allows one to quickly identify and define (i.e., categorise) the world around them. It is the process we in the West understand as becoming *an* individual. Thus FASE are crucially enacted or prevented through the visual, since that is how one seemingly 'projects' one's self outside of one's body. The sense of sight helps one maintain the borders of one's body through deciphering and mediating what can and cannot come into contact with the imagined borders of the body. As Colebrook explains, 'When the eye is privatized it takes on the mode of viewing from the point of view of the bounded and isolated individual, no longer *feeling* as part of its own life the infractions on other bodies. Instead the eye becomes a detached observing organ, intensifying the border between self and other.'[25]

Many classical and contemporary sociologists have given perception and sensation at least a cursory glance.[26] Simmel (2002) and Goffman (1963) give particular attention to sight and eye contact in social life, as does Child (1950), 'who claims that perception buttresses the sociology of knowledge, and Lowe (1982), who offers that perception is the link between the content of thought and the structure of society.'[27] The way we use our senses in daily life and what and how we perceive are culturally constructed processes based upon various 'structures of expectation'.[28] In the contemporary West, because we are inculcated in early life to restrain our sensory-embodiment and manage our physicality in socially specific ways, we learn to use our 'sense of sight' to apprehend the world most readily. Goffman highlights this in his analysis of perceptual framing and the import of the visual. He says, 'What is heard, felt or smelled attracts the eye, and it is the seeing of the source of these stimuli that allows for a quick identification and definition—a quick framing of what has occurred.'[29] For the *homo clausus* subject, vision is extremely important because it is most closely tied to rationality and enables the rationalisation of the material world. Sight is experienced as the sense most closely related to the mind (the mind's eye, as it were) and recruited to do the work, at times, of all of our senses.[30] 'Seeing is believed to be unique among the senses in terms of its ability to provide the undisputable [*sic*] truth.'[31] The way FASE are twined with sensory-monitoring is a sophisticated system of body-identity management: Learned shame and embarrassment foster distrust in one's body; fear and anxiety create distance from feelings of shame and embarrassment; and the sense of sight assuages fear and anxiety moment to moment.

Sight is arguably the most highly regarded sense in Western society. It is the primary mediator in daily life and, while maintaining the status quo, it arguably keeps us from a fuller sensory-embodied life.[32] As

Colebrook explains, 'The eye increasingly becomes a site of passional attachment in itself: if in the primitive social machine the eye operates haptically . . . the eye of modernity becomes a *reading* eye centred on man as an organism who views the world as so much calculable material.'[33] Vision and our overemphasis of it keeps us from using and being sensorially aware bodies in new and different ways because it has been socialised for purpose-oriented use, which tends to be dis-embodying, rendering flesh into passive materiality while the knowing-eye reads the world. Taken together, vision is hierarchical: valuing front-body, forward-looking, heteronormative progression; selective: according to 'optical socialisation'; and limiting of embodied innovation on an individual scale.

According to Elias (2000), as the process of civilisation and individualisation moved along, social life started to become more predictable, quieter, and less dangerous. People learned how to discipline and manage themselves in order to avoid breaking into wild and often violent swings of emotionality. Instead of social issues, these drives and passions became the domain of the individual, and restraint was expected for the sake of others.[34] Social control shifted to self-control and danger became 'internalised' instead of acted 'out'.[35] This is where the import of the visual takes hold for the development of *homo clausus* subjectivity. As outbursts and dangerous rages are rendered socially unacceptable, the individual not only learns how to restrain such urges, through a reconditioning of the embodied-self (to a dis-embodied-self), but also learns to observe, to *view* the subtlety and distinctiveness of other people's actions with a new awareness. As Elias explains, 'Just as nature now becomes, far more than earlier, a source of pleasure mediated by the eye, people too become a source of visual pleasure or conversely, of visually aroused displeasure, of different degrees of repugnance. The direct fear inspired in people by people has diminished, and the inner fear mediated through the eye and through the super-ego is rising proportionately.'[36] This situation is the outcome of processes that occurred gradually over hundreds of years, but it is now replicated within each individual, on a much more condensed scale, in the matter of a few years of one's upbringing. It can be seen, as Bourdieu describes it, as 'a structuring structure, which organizes practices and perceptions of practices.'[37] Through this individual yet collective self-regulation enacted through the eye and understood through the rational mind, the inner sense of self, from which individuals watch other individuals, is further consolidated. Foucault describes this as an 'inspecting gaze'. He says, 'There is no need for arms, physical violence, material constraints. Just a gaze. An inspecting gaze, a gaze which each individual under its weight will end by interiorising to the point that he is his own overseer, each individual thus exercising this surveillance over, and against, himself.'[38] The *homo clausus* self surveils itself and others to maintain heteronormative sameness and detects any difference, with its probing eye, as repulsive. At the daily experiential level this consolidation is apparent in how, instead of touching or engaging with something or someone directly—*physically with our flesh*—we merely *look*. Eye contact has replaced enfleshed engagement in many cases and helps solidify the ideals of stable, self-same *homo clausus* body-identity. While we may not *touch* very many people or things every day, we are constantly bombarded by images, sights which we visually consume (or not) in socially relevant ways. Bruner takes this point further by talking of 'perceptual readiness'[39] as a process that works in tandem with 'selective attention' (not wholly unlike Goffman's devices of 'framing', 'disattention' or 'inattention'), through which 'we seek out and register those details that are consistent with social expectations, while overlooking other details that are equally perceptible and "real".'[40] The *homo clausus* knowing-eye seeks out consistency in all bodies, in a self-same way, while overlooking other aspects. This is clear when women who appear masculine are forcibly ejected from public toilet spaces because they have short hair, no makeup on, and are not in particularly feminine clothing even though they are clearly female-bodied (e.g. breasts, hips, lack of facial hair). Thus *homo clausus* relies on the gaze to police bodies, to prop up social propriety, and to adhere to heteronormativity. Bodily fear, anxiety, shame, and embarrassment, without socially contingent ways of perceiving, would lose all currency.

DIS-EMBODYING THE MONAD

This 'stable,' 'self-distanced,' sensorial-individuated position of the *homo clausus* body-identity is at the core of what I term 'dis-embodiment'. While much of the *homo clausus* identity is based on controlling, ignoring, and denying the fleshy, physical, emotive body, one can never entirely disentangle from it. Instead, one creates distance from the fleshy body through imagined (i.e., conceptual) bodily borders, which are maintained by sight via optical socialisation and selective attention. This is a position of dis-embodiment; the subject is obviously always a body, but the body is not integrated into the most fundamental aspects of the sense of self.[41] This is not to say that it is favourable or even possible to experience an 'unfiltered' sensory-embodiment, but rather that we must acknowledge 'the vast amount of potentially perceivable data that is normally blocked from our awareness.'[42] Like *homo clausus* subjectivity generally, this state of dis-embodiment, facilitated through perceptive blockage, is not *naturally* occurring, but rather *naturalised.* The body is actively *rendered into* an encapsulating, passive mass via *homo clausus* subjectivity which is itself enfleshed.

Drew Leder, in *The Absent Body*, offers a keen phenomenological treatment of the *homo clausus* body-identity, but his understanding differs fundamentally from my conception of dis-embodiment, which I will explain below. While he seeks to move beyond Cartesian dualism, and is successful in much of his critique of that approach, his observations remain dis-embodied and heteronormative (i.e., in his case, white, middle-class, male). He reinforces the phenomenological *homo clausus* experience in his analysis of how, in the normal course of events, the body fades from experience or consciousness of the individual self. This understanding of embodiment nicely highlights the paradox within phenomenology, described by Shilling as 'having been interpreted by many theorists as analyses of the "lived body", of how people experience their bodies, [the work of Merleau-Ponty and others within] this tradition is actually concerned with the bodily *basis* of experience.'[43] By giving attention to the *homo clausus* body, Leder merely describes the dis-embodied subjective experience of a bounded self that inhabits a sealed body; the body is simply the basis for *conscious* (not bodily) experience. This is apparent in his description of the status quo, where the only concern is with movement and rational front-body use; e.g., walking down a long corridor in the airport to reach one's gate is typically a rational, progression-oriented action. The movement is about the destination, and during it, one's consciousness is seldom directed beyond the goal, beyond the rational into the sensory or embodied. Leder's mistake is that he believes it is his body which must get his mind to break the rational process, to go beyond the rational into the sensory. This understanding reifies rationality as that which is separate from the body. His mind, seemingly located in his skull, maintains *homo clausus* and Cartesian boundaries as he is unable to dissolve the imagined borders of his mind in order to permeate his 'non-active' body. That is to say, it is his mind, his conscious awareness that must be taught how to permeate his body, not the other way around. This training of the mind is overlooked by dis-embodied phenomenology. In Leder's formulation of embodiment, conscious awareness of the self (i.e., the active mind) is *required* for experience—whereas sensory-bodily experience or embodied awareness rarely leads to consciousness. This is because the body is not understood as active materiality but rather passive and in most cases, stable. Leder cannot bring his mental awareness into his 'motionless' body and thus he concludes that the non-active body 'disappears' from awareness *because it is not active.* He says, 'Bodily regions can disap-

pear because they are *not* the focal origin of our sensorimotor engagements but are backgrounded in the corporeal gestalt: that is, they are
for the moment relegated to a supportive role, involved in irrelevant
movement, or simply put out of play.'[44] The inherent implication in his
conclusion, laden with visual metaphor and bodily hierarchy, is that
only overtly 'relevant' movement can warrant mental attention. This is a
mind-body relationship that is focused on *doing* not on *being*. The mind
can only become aware of the body when the body is *doing* something
practical or painful, i.e., in Leder's formulation, doing something new,
different, or wrong. Otherwise, the body should 'disappear' into its normally passive state. There is no space for the mind to become aware of
the *being*, living body or for embodied awareness of the mind. There is
a clear split between the two.

In this theory, Leder reveals to his readers that, in his acceptance of
his mental processes as primary, he is unaware of the patterns of his
own mind.[45] This is as an explicitly heteronormative (and arguably,
white male) understanding of embodiment insofar as heteronormativity
is generally concerned with progression and usefulness[46] (i.e., production) and thus views queerness (as one oppositional framework to
hetero) as 'stagnant and useless'.[47] Leder's apprehension of embodiment is one of selective attention, based upon a particularly Western
style of centring attention on a 'focal point'.[48] This way of being and
comprehending (central to *homo clausus* identity) operates via a pattern
of exclusion and ignorance of sensory-embodied experience in social
settings as well as when one is alone. As Zerubavel explains, 'Ignoring
something is more than simply failing to notice it. Indeed, it is quite
often the result of some pressure to actively disregard it. Such pressure
is usually a product of social norms of attention designed to separate
what we conventionally consider "noteworthy" from what we come to
disregard as mere background "noise".'[49] Leder furthers this paradigm
of social norms (i.e., heteronormativity) in his account of the body. He
believes the body, when not engaged in purposeful action 'that creates
our environment and governs our daily routines' is not part of our
experience, because we are not consciously aware of it, but also that the
body 'can abruptly reappear as a focus of attention when we are ill or in
pain and when our bodies are at their least socially productive.'[50] Leder
terms this reappearing of the body '*dys-appearance*'. 'That is, the body
appears as thematic focus, but precisely as in a *dys* state—*dys* from the
Greek prefix signifying "bad", "hard", or "ill".'[51] He explains that 'dys-
appearance characterizes not only the limits of vital functioning but
those of affectivity' where, for example, he may experience some emotions within himself 'holding sway . . . as an alien presence' that he

cannot get rid of.[52] Leder adds that 'anxiety provides a good example of this phenomenon.'[53]

While compelling, this understanding of embodiment clearly suffers from monadic *homo clausus* experience. It fails to question how certain engagements, emotions, and ways of understanding are learned in such a way that typical bodily 'relegation' can even occur. This is evident in his use of 'focal origin' and the ideas of 'disappearance' and 'dys-appearance' in and of themselves. 'Focal origin' implies an understanding of *having* not *being* a body, which, while it may not be entirely visually based, is steeped in optic metaphor and the use of sight to notice a 'region' of the body (e.g., he cannot see his legs when he is sitting in a chair reading because his eyes are engaged in something else—reading—thus his legs fade from his experience). For Leder, experience is seemingly only possible through consciousness (i.e., rational, mental apprehension), and he is unable to assign any value to non-activity, stillness, or the back-body, all of which can be read as 'useless' or 'queer' in his formulation of body-identity. What he problematically fails to acknowledge is that his entire thesis is predicated upon a highly rationalized style of dis-embodiment, which is highly demonstrative of adult *homo clausus* body-identity. As Shilling explains, 'The body only fades for Leder when it has become sufficiently *rationalized* to be engaged in instrumental action and is actually *engaged in* such action. Thus Leder's account can usefully be read as an ethically worrying explanation of what happens to bodies when they become locations for the effects of a highly rationalized society.'[54] Leder's very limited understanding and experience of being both conscious and embodied make for a troubling ethics of personhood. His conception implicitly honours a straight, front-body, forward motion,[55] sensory-embodiment that supports the *homo clausus* way of being insofar as it implies that there can only be one main point of bodily sensing at a time; meaning the body is always separated (i.e., sensorially individuated) through rationality in a hierarchical way. His account, more than embodied, is *cognitive*; the description of 'fading' and his inability to notice or be conscious of certain sensory, perceptive, and emotive happenings is akin to a 'cognitive structure' concerned with 'relevant attributes' defined by Fiske and Taylor as 'schema'.[56] Cerulo explains that 'schemata . . . allow the brain to exclude the specific details of a new experience and retain only the generalities that liken the event to other experiences in one's past.'[57] Cognitive schemata produce sameness (normality) where there may be differential ways of being. Thus one may fail to recognise their own inherent difference in their ongoing embodiment. When the body is understood as passive materiality and only 'normal' when one is not conscious of it, experiencing only makes sense through focused rationality.

The two types of disappearance Leder terms '*focal*' and '*background*' honour a 'focused' perception that, again, is rational in its splitting of the body and sensing capacities into 'knowable regions'.[58] Where the body cannot be split and conquered rationally it becomes a mystery. The issue of 'disappearance' and 'dys-appearance' as ways to describe one's experience of the body in use inscribes a sort of mystical unknowable sense that the body on its own is something that just happens *to us* and is *naturally* out of our control. This is clear in his treatment of the 'affective' where his emotions are described as something foreign to him, which he cannot seem to master or control. Again, this is a heteronormative understanding of adult embodiment, where masculinity cannot make sense of emotionality *because* it happens *in the body*. This is clearly where Leder's phenomenology does not fulfil its mandate; this is decidedly not about the body as a living being. In his elucidation of his experience of anxiety, he does not seek to understand *why* he experiences this state (the 'why' of emotionality is not a productive question for heteronormative masculinity), or where in his body he feels it, but rather how he experiences it. That is, how he is *conscious* of it. This description is useful for creating a *rational* and highly distanced understanding of an emotion but does very little to understand why he is experiencing it in the first place, as if emotions just 'appear' on their own. This also means that he is disconnected from his emotions and thus he is creating the 'out of control' feeling in his own body that patriarchy inscribes onto women's bodies. Since women, historically, have been 'expected' and 'allowed' to be emotional (at least to a certain degree) they have a better opportunity and material history for understanding why they feel certain emotions. Thus, logically, we can draw the conclusion that women may generally be less subject to experiencing their emotions as 'out of control', since they are given the space to engage with them, to feel them, which is in contrast to Leder's experience and the general masculine pressure to disconnect from and rationalise emotional processes—rendering masculine emotionality more 'out of control', because more out of touch (i.e., they don't *feel* their emotional processes, they are disconnected) than women.

Similarly, despite the inclusion of the sensorimotor apparatus as something seemingly dispersed, there is, in Leder's work, a strong undercurrent of rational focused knowledge as the way to *access* sensing, feeling, and perception. This conception of the body is not about embodied-knowledge, it is about knowing through the focused mind (i.e., consciousness) and maintains sensorial individuation and dis-embodiment. For Leder, consciousness and experience are nearly synonymous. My conception of dis-embodiment presupposes Leder's mystical body, which is experienced as *naturally* absent (passive) or dys-appearing (active). It goes beyond this by highlighting how his account unintentionally points to the ways that the body is *made into* something we *learn* to keep from conscious awareness, not something that is naturally separate from it. Unlike the mystery of the disappearing body, dis-embodiment implies that we actively diminish our awareness and distance our consciousness from our bodies—that is, we create that sense of mystery about ourselves through an *active*, ongoing process of dis-embodiment. One implication of my approach here is that we can also actively shrink that distance, should we be inclined, in order to create more, and greater forms of, embodied awareness. Through *homo clausus* dis-embodiment the rationally bounded mind (seemingly confined to one's skull) is experienced as the central source of knowledge, information, and understanding, while the body is merely the encapsulating, pragmatic mass, requiring control and management. This is a body that one inhabits (and inhibits) through rationally embodied connections (dis-embodiment), but not understood as a body that one *always already is*. *Homo clausus* body-identity takes the 'raw data' of sensory-embodied experience and through 'consciousness' turns it into knowable information; this is the very limited, narrow depiction of rational experience. The body is always active in this formulation but rarely acknowledged as such; thus there is little room for sensory-embodied experience and virtually no space for thresholds of becoming-other. Leder's *homo clausus* phenomenology exemplifies a self who *has* a body, not a self who is thoroughly and actively embodied. Fortunately, this is not the only way to conceptualise or to experience our embodied being.

CONCLUSION: BODY AS TERRITORY

Homo clausus subjectivity is based upon a self within a body. This body-identity is not natural, but rather requires a substantial amount of work and effort to maintain. The inner sense of self is consolidated through the rational control of desires, drives, and emotions and the body is made into a conceptual case through the manifestation of imagined borders, which must be managed accordingly. This is accomplished (at least partly) through sensorial individuation, selective attention, and optical socialisation. Independently managed through socially instilled fears and anxieties of shame and embarrassment (which seek to make individuals *feel* disconnected from society), the individual body becomes the vessel that contains the all-important self. Individual consciousness, mediated through the distancing eye, becomes the primary mode of subjective experience. While we may interact with and even rely on other individuals in daily life, our self seemingly remains locked inside our body, constant and without direct social interference. The self can only be accessed through the conscious, rational mind, which one apparently has control over, as the body is not a source of experience, but merely the basis for it. For most who are raised in a Western society, it is extremely difficult to 'imagine that there could be people who do not experience themselves in this way as entirely self-sufficient individuals cut off from all other beings and things.'[59] This phenomenon, where the core or true self 'appears likewise as something divided within him [*sic*] by an invisible wall from everything outside, including every other human being' *is homo clausus*.[60] In experiential terms, it is a detached style of *inhabiting* the body, where the *individual* is the primary social entity. Rather than the body being thought of and thus experienced as active and integral to social life, it is understood as a tool, vessel, or machine required to sustain the self.[61] It works along a singular heteronormative axis of sameness.

While this may seem logical and even *natural* to some, as 'many sociological theorists[62] today accept this self-perception, and the image of the individual corresponding to it, as the untested basis of their theories', the blatant rationalisation and oversimplification of the self-body relationship is highly problematic.[63] It renders individual social identity as the way one *interacts with* but does not necessarily *constitute* society. This situation is the precursor for Elias's examination in *The Society of Individuals*, wherein he interrogates the relationship of these two 'parts' of the 'whole'. While we may think of the individual as the pre-social or constant core entity and the socialising or social conditioning a person goes through as two separate layers they 'are in fact nothing other than two different functions of people in their relations to each other, one of which cannot exist without the other.'[64] Individuals are both shaped by their relations to other people and actively shape other individuals in and through their relations. To think of this merely in terms of self-society interactions is a grossly reductionistic understanding of the social process, of embodiment, and of the self. The very existence and development of individuals is only possible through contact with many other people, other bodies that are different to one's own. If anything, embodiment *as the seat of experience* is the given, the logical, ontological starting point, and the individual self is formed from and through contact with others. Yet the body gets left out because the focus is on individual subjectivity as the primary mode of experiencing, not on embodiment. Furthermore, the potential for personal change (physical, ethical, political or otherwise) is barely possible for the *homo clausus* individual. As any process of de-territorialization is highly threatening for *homo clausus*, there are very few opportunities for personal growth, emotional maturity, for becoming-other. When they do happen they may be experienced as a particular crisis (e.g., 'identity' or 'mid-life'), an 'epiphany', or entirely devastating. In order to push Elias's conception further and to better grasp the alive, becoming body, we must seek to move experience and knowledge beyond the construction of the walled-in, highly territorialized, individual self and allow materiality to feed back into and disrupt the purely cognitive loop.

NOTES

The opening quote is from Costica Bradantan, 'Scaling the "Wall in the Head"', *New York Times Blogs, Opinionator*, November 27, 2011, http://opinionator.blogs.nytimes.com/2011/11/27/scaling-the-wall-in-the-head/.

1. Norbert Elias, *The Civilizing Process: Sociogenetic and Psychogenetic Investigations* (Oxford: Blackwell, 2000).

2. Norbert Elias, *What is Sociology?* (New York: Columbia University Press, 1978), 116.

3. Chris Shilling, *The Body and Social Theory*. 3rd ed. (London: Sage Publications, 2012).

4. Shilling, *Body and Social Theory*, 175.

5. Shilling, *Body and Social Theory*, 175–76.

6. Shilling, *Body and Social Theory*, 176.

7. In one of my upper level undergraduate courses I asked my students why they thought we counted our senses and they could easily reach rational explanations. When I asked them to describe experiences in daily life when they actually use their senses separately, they quickly realized that was much harder to rationalise. This discussion ultimately undermined their initial rational explanations and if nothing else provided some interesting and engaging bodily-awareness building.

8. Asia Friedman, 'Toward a Sociology of Perception: Sight, Sex, and Gender', *Cultural Sociology* 5, no. 2 (2011).

9. Norbert Elias, *The Society of Individuals* (London: Continuum, 1991).

10. Elias, *Civilizing Process*, 475.

11. Stephen Menell, *Norbert Elias: Civilization and the Human Self-Image* (Oxford: Blackwell, 1989), 88, my emphasis.

12. Luce Irigaray's *Speculum of the Other Woman* (1985), where she rewrites Freud and interrogates many other philosophers, is a clear case in point.

13. Emily Martin, *The Woman in the Body: A Cultural Analysis of Reproduction* (Boston: Beacon Press, 2001), 22.

14. Martin, *Woman in the Body*, 21.

15. Martin, *Woman in the Body*, 22.

16. Elias, *Civilizing Process*, 410.

17. Norbert Elias, *The Society of Individuals* (London: Continuum, 1991), 122.

18. I use this term in contrast to 'self-awareness', which points to a more thoroughly embodied, emotionally and sensorially engaged process. Self-consciousness tends to be a rational, reflective process where one projects their awareness 'out' of and back 'in' or 'on' to their body. Self-consciousness tends to serve a social function, whereas self-awareness is more personal and diffractive (happening at several levels of being), rather than merely reflective. For more on this distinction see Barad (2007).

19. Elias, *Society of Individuals*, 122.

20. Erving Goffman, 'Embarrassment and Social Organization', *American Journal of Sociology*, 62, no. 3 (1956): 264.

21. Thomas J. Scheff, 'Shame and Conformity: The Deference-Emotion System', *American Sociological Review*, 53, no. 3 (1988): 398, original emphasis.

22. Scheff, 'Shame and Conformity', 397.

23. Scheff, 'Shame and Conformity', 398, original emphasis.

24. Goffman, 'Embarrassment and Social Organization', 264.

25. Claire Colebrook, *Deleuze and the Meaning of Life* (London: Continuum, 2010), 17.

26. For example, according to Friedman ('Toward a Sociology of Perception', 189), Cerulo (2002) 'locates traces of what she calls a "sociology of sensation" in the work of Durkheim (1966 [1951], 1995 [1912]), Marx (1978), Cooley (1962 [1909]), Schutz (1951), and Weber (1946).'

27. Friedman, 'Toward a Sociology of Perception', 189.

28. Deborah Tannen, *Framing in Discourse* (Oxford: Oxford University Press, 1993).

29. Erving Goffman, *Frame Analysis: An Essay on the Organization of Experience* (Boston: Northeastern University Press, 1986), 146.

30. Consider the vast array of visual metaphors that permeate our language (e.g., 'seeing is believing'); we even use vision-based euphemisms to obfuscate the use of our other senses. For example, often when we want to touch and hold something we say 'let me *see* that'. This means that even when we are able to use our sensory-embodiment in fuller ways we still credit the rational, sense of sight for the action.

31. Friedman, 'Toward a Sociology of Perception', 189.

32. Consider how contemporary Western ideals understand those with sight 'loss' as ill-equipped for contemporary life—yet possessing an 'extraordinary' ability to engage with their 'other' senses.

33. Colebrook, *Deleuze and the Meaning of Life*, 18, original emphasis.

34. It was man's civil duty, and since women were understood as particularly emotional beings, they were not fit for public life and thus not considered citizens. This still affects citizenship and civil rights today. For a discussion of the implications of such an ethics see for example Irigaray (1996).

35. When children cannot yet control their body-selves according to social standards they are described as 'acting out' their urges and impulses.

36. Elias, *Civilizing Process*, 420.

37. Pierre Bourdieu, *Distinction: A Social Critique of the Judgement of Taste* (Cambridge, MA: Harvard University Press, 1984), 420.

38. Michel Foucault, *Madness and Civilization: A History of Insanity in the Age of Reason* (New York: Random House, 1988), 155.

39. Jerome S. Bruner, 'Social Psychology and Perception', *Readings in Social Psychology*, 3 (1958): 92–93.

40. Friedman, 'Toward a Sociology of Perception', 191.

41. In the most extreme cases, transgender and transsexual individuals feel that they are 'trapped' inside of the '*wrong*' body and take steps to change the physical/biological makeup of it through surgeries and hormonal therapy.

42. Friedman, 'Toward a Sociology of Perception', 192.

43. Chris Shilling, *The Body in Culture, Technology and Society* (London: Sage Publications, 2005), 56.

44. Drew Leder, *The Absent Body* (Chicago: University of Chicago Press, 1990), 26, original emphasis.

45. Just as one can learn to become aware of one's patterns of movement or breath, one can learn to notice patterns of thought. Learning this sort of awareness is part of very basic meditative practices which teach one to firstly notice that they are thinking; secondly notice the types of thoughts they tend to have (e.g., if they are regarding the future or past); and thirdly learn that simply noticing them can be enough, rather than judging or engaging them.

46. This is in parallel to 'Georges Bataille's work on eroticism, waste, uselessness and unrecuperability [*sic*].' Noreen Giffney, 'Queer Apocal(o)ptic/ism: The Death Drive and the Human', in *Queering the Non/human*, ed. Noreen Giffney and Myra J. Hird (Hampshire, UK: Ashgate, 2008), 59.

47. Giffney, 'Queer Apocal(o)ptic/ism', 68.

48. Richard E. Nisbett and Takahiko Masuda, 'Culture and Point of View', *Proceedings of the National Academy of Sciences of the United States of America*, 100 (2003), 11163.

49. Eviatar Zerubavel, *The Elephant in the Room: Silence and Denial in Everyday Life* (Oxford: Oxford University Press, 2006), 23.

50. Shilling, *Body in Culture*, 57.

51. Leder, *Absent Body*, 84, original emphasis.

52. Leder, *Absent Body*, 84.

53. Leder, *Absent Body*, 84.

54. Shilling, *Body in Culture*, 58, original emphasis.

55. This can be read as a heteronormative, linear, success and progression-oriented way of being.

56. Susan T. Fiske and Shelly E. Taylor, *Social Cognition: From Brains to Culture* (New York: McGraw-Hill Higher Education, 1991), 15.

57. Karen A. Cerulo, *Culture in Mind: Toward a Sociology of Culture and Cognition* (New York: Routledge, 2002), 8.

58. Leder, *Absent Body*, 26.

59. Elias, *Civilizing Process*, 472.

60. Elias, *Civilizing Process*, 472.

61. When I tell people that I do 'sociology of the body' an overwhelming percentage cannot even understand without prompting from me, how the body could possibly fit into sociology. This alone is a clear testament to the disembodied nature of daily life in the contemporary West. I can generally clear up some of the confusion by simply asking, 'well you need a body to be a member of society, right?' or 'doesn't everyone have a body?'

62. The examples Elias (*Civilizing Process*) gives are Descartes, Max Weber, Parsons, Leibniz, and Kant. Leibniz stands out among them as being credited for monadology, an important step away from egocentrism.

63. Elias, *Civilizing Process*, 472.

64. Elias, *Society of Individuals*, 60.

2

HOMINES APERTI AND POST-STRUCTURALISM

The *homo clausus* subject embodies a paradox. This subject is an individual who can only exist as such through the denial or ignorance of the (collective) social processes which help create a sense of territory, of individuality. Theories and philosophies which interrogate or engage with this subject may acknowledge the social (or at least the existence of other individuals), but the emphasis tends to be on the individual mind-self within the 'passive body'. Such accounts of the self fail to realise or neglect to give attention to the ongoing social processes undertaken (and thoroughly embodied) which *actively render* the body *into* something that *can be* rationally understood as passive, merely biological, materiality. While some philosophical approaches, such as phenomenology, claim to capture the 'lived experience', they are not always successful in moving beyond dualistic, representationalist understandings of embodied selfhood. According to Elizabeth Grosz, phenomenology is inadequate because it 'assumes the functional or experiencing body as a given rather than as the effect of processes of continual creation, movement, or individuation.'[1] For example, while Leder's (1990) phenomenology, discussed at length in the previous chapter, gives attention to the body as the *basis* for experience (but not the active source of it), it does not open up or challenge the *homo clausus* body-identity as natural, given, or occurring from *within*. Rather, it reifies that notion. In Leder's conception, as explained by Shilling, 'There is little suggestion that the body can become a major, prolonged focus of attention in its "normal" state; that it can become a sensual vehicle for creativity or an explicit site for individual development.'[2] Leder supports the Western philosophical tradition's understanding of monadic embodiment, where individuals possess an 'inside' life from which they rationally view and comprehend the 'outside' world. This binary, like most, can be easily problematised due to its simplistic, reductionistic nature.

Elias attempts this problematisation through the concept of figurations, a term chosen to highlight those webs of interdependency and interrelatedness in which all humans exist. Whereas Parsons et al. 'take the privacy and individuality of every person's bodily sensations as evidence that man is by nature in effect a self-contained and solitary being', Elias swiftly exposes the oversimple nature of this view.[3] He shows how both individuality and emotionality are directly connected to and reliant upon the social, upon other 'individuals', highlighting that we are *social* beings, not solitary ones. For example, he points out 'that each person's striving for gratification is directed towards other people from the very outset' and that the experience of gratification is not 'itself derived entirely from one's own body—it depends a great deal on other people too.'[4] This underscores how embodied experience is construed as purely individual, when in practice (i.e., daily life) it is wholly reliant upon others, and not only in evocation, but as a *possibility for experience in and of itself.* To account for this interrelatedness, and to expose its continued misconstruing by Western philosophy and sociology, Elias posits humans as *homines aperti*, or 'opened selves'. This conception sits in dialectical opposition to the *homo clausus* model of the sealed-in, entirely territorialized, independently functioning self by targeting the collective and interchanging nature of identity construction and experience.

In this conception, society is not merely a collection of closed, independent individuals, but rather open, interdependent, people; i.e., there is no 'person' in the singular without 'people' in the plural. The potential to shift from an individual, monadic experience of *homo clausus* to that of dynamic, open, *homines aperti* is more productive for understanding individuality, society, and everyday life. As Elias explains, it is 'essential if people are to recognise that the apparently real partition between self and others, the individual and society, subject and object, is in fact a reification of the socially-instilled disengagement of their own self-experience.'[5] That is to say, in order to become an individual, to develop *homo clausus* body-identity, one must manage, control, and sensorially discipline one's body. This process, as explained in the previous chapter, is socially specific. Individuals are constructed through social processes that rely on dis-embodiment, sensorial individuation, selective attention, and optical socialisation. These are processes of a representationalist epistemology through which materiality is understood as passive and stable. Likewise, in order to understand how societies consist of and function through individuals, it is vital to study how individual body-identity experience is fundamentally built upon this dis-embodiment and sensorial individuation *as socially instituted*. This 'socially instilled disengagement' has been productively explored by social constructionist and post-structuralist[6] theorists (e.g., Erving Goffman, Judith Butler, Michel Foucault) who posit an understanding of identity that can be described as largely social, that is, as *homines aperti*. Many of these approaches, including their feminist, queer, and linguistic variants, aim to open the closed *homo clausus* subject through identity politics, queerings, discourse, and fragmentation; through disrupting bodily boundaries, the stable sense of self, and the individualistic, dualistic notion of inside/outside. These re-imaginings re-present *homo clausus* as *homines aperti* and represent a move towards an understanding of society as a collection of open, interrelated people, where experience is contingent not on the individual rational mind, but on social life.

Questioning the source of the 'basic identity' (e.g., the body as 'neutral' ontology) that *homo clausus* subjectivity relies on for 'stability' enables re-conceptualisations of materiality, identity, sexuality, gender, and society. Rather than a basic identity constructed along the heteronormative dialectic of sameness coming solely from one's body and then developed through the rational mind (with little contribution from the social realm), postmodern approaches posit that identity is inherently social and cannot be made separate from social life. In what follows I outline the shift from *homo clausus* to *homines aperti* through three approaches that engage with and/or further theoretical developments which have been vital for new ways of thinking through material human life. First, I focus on Erving Goffman's and Judith Butler's approaches to body-identity as inherently connected to social performances, which are understood as explicitly discursive. Second, by focusing on Sheila Cavanagh's use of the theoretical (Lacanian) mirror to discursively construct body-identity, I show how such approaches, which render materiality into 'a kind of citationality', are severely lacking in their formulation of body-identity.[7] Third, I further elucidate the limits of language and discourse that these approaches prioritise by returning to Elias's conception of *homines aperti*. Finally, I briefly introduce the alternatives I work with in chapter 3.

FRAGMENTING THE MONAD

Homo clausus subjectivity is based on an understanding of the self as an internal, initial, and persistent individualism. Alternatively, and though their ideas are not necessarily congruent, Erving Goffman's and Judith Butler's work on the social performance of gender as the basis of body-identity is focused on an external or social individualism. Put simply, Goffman's (1990, 1963) view 'posits a self which assumes and exchanges various "roles" within the complex social expectations of the "game" of modern life', while Butler's conception of the self is that it 'is not only irretrievably "outside", constituted in social discourse, but that the ascription of interiority is itself a publicly regulated and sanctioned form of essence fabrication.'[8] Butler develops Goffman's ideas and opens pathways into useful critiques of both *homo clausus* and *homines aperti*. I will deal with these theorists in turn.

Goffman's approach, while maintaining some of the 'basic identity' traits of *homo clausus*, recognises that social life and social interaction are vital to the construction, experience, and performance of body-identity in everyday life. For him experience is not purely an individual, inner phenomenon, but instead socially contingent. He introduces a nascent sense of fluid identity (which queer and posthumanist theorists develop much further, see, e.g., Butler, 1999, 1993; Creed, 1995; Feinberg, 1996, 1998; Halberstam, 1998; Wilchins, 2002) where, rather than being constant, it can change depending on with whom and where one is engaged in social interaction; though it is implied in Goffman's work that the social roles we employ are relatively stable as is the body-self that employs them (hence the assumption of stable vocabularies of bodily idiom that underpin what counts as valid performances, back-regions, stigmas, etc.). While we can engage in different roles at different times, his understanding does not question how and where these roles originate and then become the norm. Like *homo clausus* subjectivity, Goffman's 'actors' (he was a fan of theatre metaphors) have a body-identity that is seemingly self-same, stable and heteronormative, *from* which they engage in different social roles. This is similar to a performer putting on different costumes to signal a character change—the performer remains the same while the role being played differs. For Goffman, the roles of individuals are seemingly based on and derived from the stable self inside the passive fleshy body; what changes is the setting, the staging, the interaction. Goffman's interactive theory begins to theorise the power of social interactions but does little to speak to how interaction (i.e., sociality) actually spurs the change of roles one plays. Part of the inability to account for the dynamism of interaction and the power for interaction to change one's 'role' is the failure to engage the materiality of the body at the same level as, or better still, as part and parcel of this interactive self. This theory remains rational, insofar as people are understood to *consciously* understand and actively *choose* the different roles played. Though the rationalisation of choice of role often seems retrospective, reflective, and inactive. This experience, while it may be concerned with the social, is still highly rigid, individual, and rationally directed.

Butler's (1988, 1993, 1999) approach takes Goffman's performing actors to the extreme. Her conception of performativity posits that it is *the performance*, and its constant repetition, that creates one's body *and* one's identity. She seeks to render the flesh and identity completely open to social intervention, but maintains the inside/outside binary in her description. Her primary interrogation is of conceptions of identity based upon some inner essence that is derived from the ontologically self-same material body. Her theory of performativity posits that through our social expectation and anticipation for the manifestation of some pre-social essence, which would reveal the 'nature' of one's sex via one's gender, we end up producing the very phenomenon we expect.[9] In the preface to the 1999 edition of *Gender Trouble* (originally published in 1990) she explains:

> In the first instance, then, the performativity of gender revolves around this metalepsis, the way in which the anticipation of a gendered essence produces that which it posits as outside itself. Secondly, performativity is not a singular act, but a repetition and a ritual, which achieves its effects through its naturalization in the context of a body, understood, in part, as a culturally sustained temporal duration.[10]

As individuals are vitally sexed/gendered from birth—*homo clausus* ways of being are reliant upon this performativity, and Butler seeks to expose how 'the regulatory norms of "sex" work in a performative fashion to constitute the materiality of bodies and, more specifically, to materialize the body's sex, to materialize sexual difference in the service of the consolidation of the heterosexual imperative.'[11] Sexual difference for Butler is merely biological, not ontological, and is thus restricted to heteronormative ideals. In this formulation 'the fixity of the body, its contours, its movements' are entirely material, but it is a materiality that is 'the effect of power, as power's most productive effect.'[12] This materiality is not active but *re*active: a product of the social.

For Butler there is no individual without the social. Heteronormative power structures work to manifest not only sex, gender, and sexuality, but the body *itself*. The only given in this account of dis-embodiment is power. Rather than based upon the rational mind, individual experience for Butler is entirely reliant upon the discursive practices (the performances), instituted from outside of the body through dominant power structures. This performative understanding of power-embodied is not wholly unlike the power dynamic necessary for the construction of *homo clausus* in the first place. The major difference between the two approaches (*homo clausus* and *homines aperti*) is that for Butler the body is seemingly territorialized by culture, not the individual rational mind, and thus the individual rational mind (i.e., the self) is left feeling entirely *de-territorialized*—that is, entirely open to social control.

In an effort to resituate agency and instigate opportunities for change, 'in place of the flawed conceptions of construction that circulate in feminist theory and elsewhere',[13] Butler calls for 'a return to the notion of matter';[14] though she isn't entirely successful in articulating how one goes about expressing agency from the materiality of the body into the power of discourse. This is because for her the self still isn't enfleshed (matter still isn't active here), but is both inside of or outside of the body—fragmented and over-stimulated through social power and patriarchal capitalism. Which may explain her move to conclude that the body itself must also be constructed through social power—that way even if something *seems* essential we can take refuge in knowing it isn't possible since the vessel holding that essentiality is not a 'real' given in the first place. This is an attempt to shift power from the social to the personal (but is arguably unsuccessful). For example, she usefully highlights that while gender cannot be true or false, real or fake, 'Performing one's gender wrong initiates a set of punishments both obvious and indirect, and performing it well provides the reassurance that there is an essentialism of gender identity after all.'[15] Put another way, performative body-identity is constructed *through* the social, which also constructs the body. As people engage in the same seemingly natural performances every day, based on the perceived sex of their body, their performances *naturalise* gendered body-identity. For Butler, the body is always gendered by the social, and the material makeup of the fleshy body is not prior to or more important than that gender, but is fundamental to identity. Through this conceptualisation she is able to pull 'natural gender' apart from 'biological sex', since gender comes from 'outside' (of the body, of one's sex) as does the materiality of one's sex; i.e., society constructs gender identity, which constructs the sexed-body. Thus the materiality of the body is passive, in need of socio-cultural shaping, and sexual difference is not an *ontological* difference but a tool of power. Furthermore, she points out that the need to continually reassert body-identity, through repetitive gender performance, exposes its inherent weakness, as there are always slips and errors in repetition.

Regardless of this astute observation, that repetition inherently exposes body-identity instability, people are still subject to conforming their performances to social norms. This is the only way to avoid the personal experiences of fear, anxiety, shame, and embarrassment which are still so deeply rooted in the construction of individuality itself—both for *homo clausus* individuals and those who have been rendered 'open' and/or 'fragmented' through society. While performativity and social roles are useful in *describing* and critiquing what one may witness or even experience in daily life, they are unable to dramatically challenge the idea that the experiencing body-self can be anything other than a reflection of deeply entrenched social norms according to the reproduction of monadic, individualistic *homo clausus*. As Barad points out, 'While Butler correctly calls for the recognition of matter's historicity, ironically, she seems to assume that it is ultimately derived (yet again) from the agency of language or culture. She fails to recognise matter's dynamism.'[16] Both 'social roles' and 'performative repetition', as ways of understanding the self-body, flatten embodied experience and glaze over experiential, material difference as the potential source of power.[17]

'To restrict power's productivity to the limited domain of the social, for example, or to figure matter as merely an end product rather than an active factor in further materializations is to cheat matter out of the fullness of its capacity.'[18] Furthermore, the very possibility for differential being is missed in Butler's understanding of repetition as the production of the self. Performativity as constant repetition ignores the ongoing historicity of the body and replicates the *homo clausus* tendency to continually manage the body, this time from the 'outside'. That is to say it misses that the body is always already active, entangled in performative practices and with capacities for memory. As Grosz explains,

> Life is temporal, durational, which means that within it, there can never be any real repetition but only continual invention insofar as the living carry the past along with the present. This situation implies that even a formally identical state can be differentiated from its earlier instantiations because of the persistence of memory, the inherence and accumulation of 'repetitions' in the present.[19]

Performativity treats the body as a blank slate, a passive piece of matter without any intelligence. Thus, by failing to give power to sensory-embodiment these approaches miss out on the opportunity for change and limit the ability to recognise thresholds for differential experiencing. Ultimately, these approaches do their work at the expense of the materiality of the body, as the self is still rationally contained within the de-territorialized and socially controlled body; it is this 'postmodern tendency to "textualize" or "idealize" the body in ways that ignore the "facticity" of the body . . . [that] result in disembodied accounts of social interactions and practices.'[20]

As in Leder's phenomenology, the body in these approaches remains passive flesh. The main difference between the aforementioned approaches and *homo clausus* is that in postmodern approaches sensory-embodiment is understood to be entirely open to yet constrained by social discourse instead of the rational mind. Thus, in these respects, postmodern and phenomenological approaches are not that dissimilar, especially since, as shown above, they both reveal that *homo clausus* identity is not natural, but rather naturalised through social practices. In both instances individuality is never individual. Postmodern subjects are still dis-embodied, sensorial individuated, and maintain an inner and outer self that is regulated through the visual and is still vulnerable to *homo clausus* fear, anxiety, shame, and embarrassment (as I deal with specifically in the first empirical chapter, on normative uses of public toilets). While these subjects may allow the social *into* their identity construction, they still remain singular entities that rationally inhabit a body which is thought to be non-existent without social discourse. These approaches fail to articulate how the materiality and historicity of the body can feed back into the social and discursive processes which shape them. This results in an entirely de-territorialized body which can only be (re-)territorialized by the social. The sense of self comes from the outside and is moved into the body, while the body is a marker, a symbol, a representation, or something to actively transgress or shape; it is discourse, and not the source of all thinking, knowing, and understanding. This body-identity, at the whim of social discourse, cannot readily experience the de-territorializing effects of becoming-other, of differential ways of being, because those opportunities are used instead to *territorialize*, to build one's identity. This furthers a representationalist logic of matter as passive and rigidly bound but open to fragmentation by a fickle world.

While postmodern linguistic and discourse-based approaches seek to render open subjectivities (e.g., open to influence, to change, to fluidity, to new identities), the dis-embodied position of the *homo clausus* identity remains intact as the ontological *basis* for such possibilities and instead the open/closed, inside/outside body-identity is *fragmented*. This is akin to shattering a glass window instead of sliding it open (and then using the shards of glass to continually rebuild the shattered pane).

The social conscripts the self into anxious activity on and through the passive body.

As explored above, these approaches fail to bring together materiality, rationality, and discourse in a non-dualistic, non-hierarchal, and non-chronological way and thus severely limit the possibilities for *homines aperti* subjectivities. That is to say, they rely on the individual monadic position from which queering, disconnection, and fragmentation can happen, meaning these approaches do not problematise how bodies are rendered into the vessel that contains the individual self. Instead they seek to open the already individuated body-identity (i.e., *homo clausus*) through fragmenting the 'natural' monadism via social discourse. These *homines aperti* individuals remain dis-embodied because the materiality of the body remains the ahistorical basis of the self. Part of their problem in reproducing body-selves that are merely representational and not thoroughly material is that they fail to interrogate the fact that they only work 'cognitively and visually' through 'presumptions about the transparency and accuracy of visual perception.'[21] A vital point that I explore below.

SYMBOLISATION AND FRAGMENTATION

As Barad keenly asserts, 'it seems that at every turn lately every "thing"—even materiality—is turned into a matter of language or some other form of cultural representation.'[22] This observation is at the heart of social constructionist and post-structuralist approaches to the self, as they are unable to directly embed the *living* body into their understanding of the self. The body is 'oppositely' approached (from the outside), yet still acts as the bridge between the self and the social. This is evident in how 'Butler's theory ultimately re-inscribes matter as a passive product of discursive practices rather than as an active agent participating in the very process of materialization.'[23] This power given to discourse, as introduced above, will be given a closer look in the remainder of this chapter through a critique of Canadian sociologist Sheila Cavanagh's use of the conceptual mirror in *Queering Bathrooms*.

The idea that theoretical mirrors are vital for the construction of subjectivity is rooted in psychoanalysis but continues to permeate many disciplines, including many of those working specifically with gender, sexuality, and the body. For Cavanagh, social and literal 'mirrors' discursively give relief to the self through the overemphasis of the visual, of sight—that is, by seeing, and by being seen. This is an example and reproduction of how 'the structure of watching and being watched is key to the operation of patriarchal society.'[24] Thus how someone looks, in both senses, is assumed to enable access to someone's 'inner' or 'private' life. Colebrook explains this in regard to desire:

> The privatization of the eye in late capitalism also seems to run alongside what Deleuze and Guattari refer to as a transition from the regime of the signifier—in which the viewed world is the sign of some ultimate reality—to the passional regime. . . . And Judith Butler (2005) notes, following Laplanche, the eye (now) sees its world and the image of its own ego-self as if from the point of view of an other whose desire would grant the truth of its being. I am a self only if I am recognised by an other. In this passional regime the self is created, located and privatized less by submission to a system of signifiers through which it articulates its desire than by relation to the face of an other whose desire is essentially hidden.[25]

In patriarchal capitalist society, we build our identities based on heteronormative ideals which presume desire—they inscribe it into our ways of being so it can remain hidden from view. Desire which is visible is 'different', non-normative, abject, as normative (non)desire is expected to be mirrored or mediated through capitalist consumption.

Like Butler's performativity, mirrors construct the individual from the outside-in. Thus the visual, for *homines aperti*, is highly important not just for maintaining the borders of one's body, but for the articulation and reading of identity. According to the mirror theory, individuals construct and know the borders of their bodies *and* their identities through contact with reflective surfaces and others whose desire is at once needed for affirmation while remaining inaccessible. There is less emphasis on innate truth of self and more need to be seen by others, to be recognised in order to have a sense of self at all. This understanding is based on using what is outside of the body to understand what one 'is' and what one is not on the inside. Feeling and desire is not rooted in sensory-embodiment, but rather accessed and processed through the visual. This is emblematic of postmodernity where 'the eye neither feels collectively (haptically), nor reads the world digitally (as equivalent units) but is dominated by passion and affect.'[26] While the theoretical mirror, as the mode of identity construction, fundamentally differs from *homo clausus*, by working from the outside-in, it does not disrupt individual and independent bodily boundaries. Both *homo clausus* and discursive approaches take the separate individual as the starting point.[27] Like Goffman's frame, the mirror concept is only viable through visual and rational understanding of individuality, as it is based upon 'a clear depiction of "in" and "out", it is a binary representation in which the attended and disattended are fully separated and spatially contiguous, rather than interwoven in the same conceptual space.'[28] Put simply, the power of a mirror, whether it is a reflective surface (e.g., a shop window), a sensory experience (e.g., hearing an echo), or an Other (e.g., another person, or another person who is conceived of as *different*, opposite, negatively dissimilar), to reflect back onto the borders of one's body, exposing differentness or sameness, is only viable because of disembodied identity and clear conceptions of categorisation (e.g., inside and outside). In order to work, a mirror requires the internal monadic experience of *homo clausus* paired with the external experience of sociality. *Homines aperti*, as conceived through mirrors, are only open to discursively confirming or further fragmenting bodily boundaries. Reflections are momentarily de-/re-territorializing. Self-same reflections help repetitively re-territorialize while incompatible reflections are suspect and threatening. They are negatively de-territorializing. Consider that reflection is a 'pervasive trope for knowing', a knowing that is attuned to sameness and repetition, not creation and difference.[29] Reflection thus enables the creation of types and taxonomies, a system of superficial difference that ignores the ongoing difference of all bodies and is only possible through division, simplification, and reduction.

While Cavanagh (2010), whose empirical research process and data are in some respects similar to my own, seeks to 'queer bathrooms', her project remains heavily steeped in this concept of the mirror and thus seeks to queer discourse through more *representationalist* discourse. She refers to many different mirrors throughout the text. Some are specifically sensorial (e.g., acoustic) and others more general (e.g., reflective surfaces, heteronormative people). She says, 'People in the bathroom behave like mirrors; architectural shapes and material objects, including reflective glass and metallic surfaces, all act as mirrors.'[30] In line with *homo clausus* sensing, she explains that acoustics also work as mirrors because 'Sight and sound work in concord. We learn to "see". . . with our ears.'[31] What she means, I believe, is that we must learn to deploy our other senses to manage and monitor the borders of our and other bodies in the same way we rely on our sight to make the world knowable. When enclosed in a cubicle within a toilet space, one's vision becomes less valuable, as there is little to *see*, so one must 'see' through one's ears. Characteristic of *homo clausus* understanding of the ability for sight to provide access to knowledge via the rational mind, she interprets the use of 'other' senses within this framework. Hearing is less valued than sight as a direct and reliable source of knowledge, so she couches that sensation in terms of sight in an attempt to elevate it to the all-important level of rationality. Cavanagh seeks to use mirrors to expose how public toilets are inherently discriminating to non-heteronormative folk by constantly reasserting the importance of the visual to individual experiences of public toilets. While perhaps useful for discussing the power of discourse, this, as an understanding of queer embodiment, is fundamentally flawed insofar as it supports rational, heteronormative *homo clausus* ways of being because it assumes that through the visual we have ultimate access to knowledge, experience, and understanding. As Barad explains, 'Representationalism [is] the belief that words, concepts, ideas, and the like accurately reflect or mirror the things to which they refer.'[32] Mirrors do not socially construct bodies, 'what is social is the process of selectively emphasising and mentally weighing the existing bodily similarities and differences' which 'highlights the socially constructed character of [identity] categories.'[33] The importance of the visual supports a socially constructed materiality, as it is tied directly to *homo clausus* patterns of selective attention and optical socialisation. As Arthur Frank simply states, 'the mirroring body tells itself in its image, and this image comes from elsewhere. The images this body mirrors come most often from popular culture, where image is reality.'[34] Cavanagh ultimately reifies the individual subject as existing within the confines of the body through her reliance on the social to reflect and represent 'reality' through the outside of the body. Here dominant culture determines reality. Thus embodied experience for Cavanagh is discursive, not *living* and very limited to experiences of becoming-other.

In attempting to make space for queer bodies, Cavanagh does very little *queering* (opening, twisting, disturbing) of social norms or understandings of power, but instead solidifies the need for queer bodies to become more *visible* (so that the mirrors they contact are more like them, and thus less alarming or exposing). It seems her approach is to try to disrupt people's perceptual and cognitive practices through a sort of re-optical socialisation. The logic is if people see new bodies, they will eventually accept them as one version of normal. This does very little to disrupt heteronormative *homo clausus* ways of being and is ultimately just another way to continue the cycle of FASE. Put simply, this is the use of a patriarchal system to normalise queer bodies. It is not a queering of bathrooms but instead a project to make queer bodies more visible. As Phelan astutely points out, 'While there is a deeply ethical appeal in the desire for a more inclusive representational landscape and certainly under-represented communities can be empowered by an enhanced visibility, the terms of this visibility often enervate the putative power of these identities.'[35] Furthermore, Cavanagh's understanding that public toilets are heteronormative spaces that need to be made more inclusive for queer bodies is based on the assumption that they are only experienced as oppressive, dangerous, and uncomfortable via individual and independent queer identity. As explored throughout this book these spaces are not merely difficult and problematic for those who do not subscribe to heteronormativity. Instead, it is my argument that they are heteronormatising spaces because they rely on *homo clausus* fear, anxiety, shame and embarrassment as primary states of being for individual identity. That is to say, they are oppressive for all bodies regardless of desire or identity. Therefore the visibility she is advocating risks three problematic phenomena: one, a binary understanding of queer identity versus heteronormative identity; two, a reductionistic view of the body as passive materiality, i.e., the body is merely the sum of its parts; and three, an easier inculcation into social norms once visible.

This sort of reductionism caused many queer and feminist theorists to turn away from the material body initially and instead honour the linguistic and discursive over the embodied self, and here Cavanagh is reproducing that representationalist mistrust in matter, making her approach to queerness dis-embodied. The desire for queer visibility seeks to stabilise queer identity by making it *clearly visible*. It is the use of rational discourse to empower more discourse. Put simply, this is not very queer. In many ways, stability through visibility is the antithesis of queer. It is not surprising then that Cavanagh's ultimate goal is the redesigning of public toilet spaces, not the challenging of how we understand non-heteronormative identities; though it remains unclear if that goal would enable greater visibility of difference, of *Otherness*, or better hide it.

Cavanagh exposes and renders her subjects utterly vulnerable through the inscription of the dis-embodied discourse of the mirror. The use of mirrors as a primary mode of understanding the self maintains, through the visual, the mind as the primary mode of experience. It reifies the inner self and the external world, but it is also a robust, albeit problematic example of how *homines aperti* work through interrelation and reflection. In the contemporary West, mirrors, like language do not create or give relief to the material body, they further dis-embodiment by visually condensing the sense of outer bodily boundaries and leave the inner sense of self intact yet entirely open to fragmentation through dissimilar mirrors. This is how reflective, representational, symbolic—that is, *discursive*—systems of understanding completely leave out the materiality of the body and render it into mere passive biology. These approaches work by denying the ontological basis of being by rendering the body into a *thing* through representationalist ways of being based upon an epistemology which is deeply sceptical of matter. As Barad posits:

Is not, after all, the common-sense view of representationalism—the belief that representations serve a mediating function between knower and known—that displays a deep mistrust of matter, holding it off at a distance, figuring it as passive, immutable, and mute, in need of that mark of an external force like culture or history to complete it? Indeed, that representationalist belief in the power of words to mirror preexisting phenomena is the metaphysical substrate that supports social constructivist, as well as traditional realist beliefs, perpetuating the endless recycling of untenable options.[36]

In order to better get at how *materiality* is caught up in figurational social systems and daily phenomena, such as sex-gender-sexuality and the family, there needs to be a joining of the epistemological and the ontological which honours the material body as not merely passive matter collectively produced and shaped, yet individually experienced. There needs to be a new approach to matter not the visual. While useful, these linguistic approaches are not adequate in and of themselves.

CONCLUSION: LIMITING THE LANGUAGE

This precarious situation that posthumanisms and material feminisms are working to exploit (as I explore in the following chapter), in an effort to redefine the role of matter, has not occurred without warning. Nietzsche in the nineteenth century 'warned against the mistaken tendency to take grammar too seriously: allowing linguistic structure to shape or determine our understanding of the world, believing that the subject and predicate structure of language reflects a prior ontological reality of substance and attribute.'[37] It is through this awareness, of language as metaphor for ontology, that Nietzsche is able to swiftly dismantle Descartes' dualism as a product of the way his belief in and use of language has been shaped. Descartes' dualistic understanding and elucidation of the 'thinking self' is too simplistic; rendering the mind separate from the body, even though the body is most certainly required for thinking and the existence of the mind. It is unsurprising then, as we are still suffering the repercussions of Descartes' unaware and facile treatment of the body-identity, that as *homo clausus* and *homines aperti* we make our language and discourse in the image of ourselves and we make ourselves in the image of our language and discourse. Like individuality and the visual, language is employed as though it too is a given, perfectly apt to *mirror* 'the underlying structure of the world' and the seemingly 'preexisting' phenomena that support

representationalist, realist, and social constructivist[38] beliefs.[39] As elucidated above, the linguistic and discursive are understood to impact
upon and even *produce* bodies, rendering a *homines aperti* society; but
the body, as given, stable, and/or passive, is unable to feed back into the
relationship in any way. The opportunities for becoming-other are few
and far between. These body-identities may be open, but they are not
active, instead they are anxious. Similarly, as explored in the discussion
of the mirror, the attempt to account for difference, in the form of
queer bodies, through representationalist frameworks only reproduces
heteronormative conditions for understanding. Thus rendering difference into sameness. In these attempts there is no space for the body (let
alone the development of sensory-embodied-identities) because materiality has been flattened by the power given to words, discourses, and
epistemologies that maintain the two-dimensional nature of *homo clausus* identity.

According to French sociologist and urban theorist Henri Lefebvre,
'Western philosophy has *betrayed* the body; it has actively participated
in the great process of metaphorisation that has *abandoned* the body;
and it has *denied* the body. The living body, being at once "subject" and
"object", cannot tolerate such conceptual division, and consequently
philosophical concepts fall into the category of the "sign of non-
body".'[40] Without resolving these fundamental fissures there is no way
to reconcile the internal self with the external world, even if that self is
seemingly open. These oversimplified binary constructions of body-
identity need to be brought into dialogue, need to be recognised in
their ongoing entanglement; rather than materiality existing outside of,
or in opposition to discourse, we need to productively recognise their
inextricable relationship. It is only then that we can recognise the possibilities and potentialities available for an *experiential* (not merely discursive) shift to *corpus infinitum*, an approach that understands sensory-embodiment as actively alive and part and parcel of all knowledge,
experience, and understanding.

NOTES

1. Elizabeth Grosz, *Becoming Undone: Darwinian Reflections on Life, Politics, and Art* (Durham, NC: Duke University Press, 2011), 28.

2. Chris Shilling, *The Body in Culture, Technology and Society* (London:
Sage Publications, 2005), 59.

3. Norbert Elias, *What Is Sociology?* (New York: Columbia University
Press, 1978), 134.

4. Elias, *What Is Sociology?*, 134–35.

5. Elias, *What Is Sociology?*, 122.

6. For ease of articulation I refer to social constructionist and post-structuralist approaches collectively as postmodern.

7. Judith Butler, *Bodies That Matter: On the Discursive Limits of 'Sex'* (New York: Routledge, 1993), 15.

8. Judith Butler, 'Performative Acts and Gender Constitution: An Essay in Phenomenology and Feminist Theory', *Theatre Journal*, 40, no. 4 (1988): 528.

9. Judith Butler, *Gender Trouble: Feminism and the Subversion of Identity* (New York: Routledge, 1999), xiv.

10. Butler, *Gender Trouble*, xiv–xv.

11. Butler, *Bodies That Matter*, xii.

12. Butler, *Bodies That Matter*, xii.

13. Karen Barad, *Meeting the Universe Halfway: Quantum Physics and the Entanglement of Matter and Meaning* (Durham, NC: Duke University Press Books, 2007), 64.

14. Butler, *Bodies That Matter*, 9.

15. Butler, 'Performative Acts and Gender Constitution', 528.

16. Barad, *Meeting the Universe*, 64.

17. This flattening of the experiential is the same (ignored) process that renders 'social structures' in place of practices of action, movement, thinking, and ways of being.

18. Barad, *Meeting the Universe*, 66.

19. Grosz, *Becoming Undone*, 31–32.

20. Asia Friedman, 'Toward a Sociology of Perception: Sight, Sex, and Gender', *Cultural Sociology* 5, no. 2 (2011): 195.

21. Friedman, 'Sociology of Perception', 201.

22. Karen Barad, 'Posthumanist Performativity', in *Material Feminisms*, ed. Stacey Alaimo and Susan Hekman (Bloomington: Indiana University Press, 2008), 120.

23. Barad, 'Posthumanist Performativity', 151.

24. Sarah Moore and Simon Breeze, 'Spaces of Male Fear: The Sexual Politics of Being Watched', *British Journal of Criminology*, advance access published August 9, 2012, doi:10.1093/bjc/azs033: 2.

25. Claire Colebrook, *Deleuze and the Meaning of Life* (London: Continuum, 2010), 17–18.

26. Colebrook, *Deleuze and the Meaning of Life*, 18.

27. Norbert Elias, *The Civilizing Process: Sociogenetic and Psychogenetic Investigations* (Oxford: Blackwell, 2000), 474.

28. Friedman, 'Sociology of Perception', 193.

29. Barad, *Meeting the Universe*, 72.

30. Sheila Cavanagh, *Queering Bathrooms: Gender, Sexuality, and the Hygienic Imagination* (Toronto: University of Toronto Press, 2010), 84.

31. Cavanagh, *Queering Bathrooms*, 110.

32. Barad, *Meeting the Universe*, 86.

33. Friedman, 'Sociology of Perception', 200.

34. Arthur Frank, *The Wounded Storyteller: Body, Illness, and Ethics* (Chicago: University of Chicago Press, 1995), 45.

35. Peggy Phelan, *Unmarked: The Politics of Performance* (London: Routledge, 1993), 7.

36. Barad, *Meeting the Universe*, 133.

37. Barad, 'Posthumanist Performativity', 121.

38. Feminist and queer theorists have expressed dissatisfaction at both social constructivism and social constructionism. See, e.g., Haraway 1988 and Butler 1999.

39. Barad, 'Posthumanist Performativity', 121.

40. Henri Lefebvre, *The Production of Space*, trans. Donald Nicholson-Smith (Oxford: Blackwell, 1991), 407, original emphasis.

3

CORPUS INFINITUM AND POSTHUMANISM

Problematising the *homo clausus* (person closed) through its undeniable interrelatedness to other people, or *homines aperti* (persons opened), enables us to expose the opened/closed, inner/outer, same/different binaries, inherent in both of these conceptions of subjectivity. In doing so, we can locate the limits of these body-identity models, both of which severely ignore or limit sensory-embodiment and instead rely on specific rational processes. Through positing body-identity as primarily enfleshed, sensorial, and undeniably entangled (materially *and* discursively), I propose the term '*corpus infinitum*' to push Elias's project further. Both '*homo clausus*' and '*homines aperti*' take individual *selves* as given, as the primary avenue for understanding societies, while also minimising or entirely neglecting the role of sensory-embodiment. This somewhat surprising error, considering Elias's keen attention to the formulation of the self as socially situated, in which sensory-embodiment is crucially overlooked, is what I seek to rectify theoretically and experientially throughout this book. Elias's *homines aperti*, while a move towards a better understanding of living bodies, suffers from the social constructionist tendency to rely on atomised social categories and binaries, for example, inside/outside, mind/body, self/other. This representationalism extends far beyond bodies and into how we conceive of the world generally. As Karen Barad explains, 'Representationalism takes the notion of separation as foundational. It separates the world into the ontologically distinct domains of words and things, leaving itself with their linkage such that knowledge is possible.'[1] This separation requires that things, subjects, objects are contained to their own matter. For example, bodies have boundaries that are sealed and stable. That means matter is not active but rather passive and in need of inscription or shaping or some such ongoing management. This approach to the world generally, and the body specifically, renders an insurmountable rift between things, subjects, objects, and language. This is an epistemology based on atomisation, and to borrow Kristin Ross's language, we must recognise our 'radical atomisation' before any change is possible.[2] That is, we must re-place matter as alive, as active at the centre of our understanding.

Post-structuralist and social constructionist approaches, while generally very useful for identifying the normative perceptual and cognitive practices (schema) we employ (e.g., I see a person who is wearing a dress so this person must be a woman), they are not successful in identifying the sensory-embodied practices that we could potentially engage because they rely too heavily on the visual-mind as the source of/for experience, knowledge, and understanding based on categories of same/different. Via *homo clausus* individuality we learn to perceptually identify and categorise sameness and difference in a simplistic two-dimensional way (e.g., alike/unlike); this, for example, is the only way that the mirror concept could be productively employed, since it works through reflection of sameness/difference. Furthermore, reflection (or a reflexivity as a methodology) 'invites the illusion of essential, fixed position' from which sameness/difference can be ascertained.[3] This binary is perhaps the most pervasive dualism inherent to *homo clausus* individuality. It is, in many ways, the same dualism as 'inner/outer' in terms of constructing and maintaining the borders of the body, and one that I hope to dislodge from conceptions of the self through my introduction of *corpus infinitum*, the boundless, unlimited, indefinite body-self.

Corpus infinitum entangles identity with *fleshy living bodies* as fundamental for social life generally, and crucial for the conscious rational self specifically. This signals a move beyond the modern independent self and postmodern interdependent selves to a posthumanist-materialist understanding of the self as always already entangled in becoming phenomena, or happening *through being*. For *corpus infinitum*, consciousness is not merely mental, visual, or rational, but fully living sensory engagement. It is attentive to *patterns* of difference, not simply representations of sameness; it is not reflective but instead diffractive.[4] As Barad explains, 'Donna Haraway proposes diffraction as an alternative to the well-worn metaphor of reflection. . . . Diffraction can serve as a useful counterpoint to reflection: both are optical phenomena, but whereas reflection is about mirroring and sameness, diffraction attends to patterns of difference.'[5] In this conceptualisation of subjectivity, sensory-embodied experience is never stable or bounded—that is, concerned with maintaining and replicating sameness of being and experience—because it recognises the active, open-ended, and ongoing nature of material being and engagement. It is through the active, ever-changing, differential experiences of materiality that consciousness, knowing, and understanding can happen. Diffraction as a vital way of elucidating *corpus infinitum* is also the methodological approach I employ as discussed in the introduction. Diffraction requires more subtle and developed sensory engagement; it necessitates 'the processing of *small but consequential* differences' and 'the processing of differences . . . is about ways of life.'[6] Thus *corpus infinitum* is not focused on sameness or difference, inner or outer, because it is less invested in classificatory systems and more concerned with understanding how sensory-embodiment is minimised through such systems. Furthermore, it takes difference as its condition for being, not some monstrous thing to avoid, because living sensory-embodiment is always already an ongoing process. It is never self-same. Ultimately, by recognising how, where, and when we're limited in daily life, we can begin to recognise how such minimisation can be expanded to enable fuller experiences, to seize the thresholds of becoming already inherent to, yet overlooked, in our daily 'ways of life'.

In what follows, then, I focus on material feminist and posthumanist philosophies, which work to identify and productively correct the binaries, biases, and material losses that are essential to the preceding approaches. To give life to *corpus infinitum* subjectivities, I unpack and connect these theories in three entangled ways. First, I show how the preceding approaches highlight the need for the joining of ontology and epistemology (i.e., ways of being and systems of knowing and/or classification), for example, through Barad's onto-epistemological re-appropriation of Butler's performativity. Second, I put forth the entanglement sex-gender-sexuality, to illustrate how onto-epistemology and material discourse operate experientially and theoretically, in the construction of body-identity experience. Third, I highlight the importance of reading Elias through a posthumanist-materialist lens. In doing so, I reveal the possibilities and potentialities intrinsic to the re-conceptualisation of living sensory-embodiment as vital for not only the construction of the self but for all experiencing, knowledge, and understanding—a situation which maintains the personal yet aims to move beyond the (e.g., patriarchal, capitalist) individualism that animates the preceding approaches. Within the previous two chapters I began to point to the ways that practices of perception are used in systems of knowledge as important to *homo clausus* subjectivity. This is an important illustration of an onto-epistemology (albeit a unidirectional, and overly rational one) and a vital point of convergence in beginning to elucidate *corpus infinitum*.

MAKING CONNECTIONS

Karen Barad poses a vital question when asking 'How did language come to be more trustworthy than matter?'[7] Language, existing in the realm of the rational, has not only been opposed to the material body per se, but worse, it has been allowed, encouraged even, to usurp it entirely. This is evident at least in Goffman's, Butler's, and Cavanagh's failed attempts to flesh out the material, as they remain firmly planted in the discursive. For example, as Barad explains, 'it is not at all clear that Butler [following Foucault] succeeds in bringing the discursive and the material into closer proximity. . . . Questions about the material nature of discursive practices seem to hang in the air like the persistent smile of a Cheshire cat.'[8] Postmodern accounts that believe the body is materialised *through* discourse cannot account for the materiality of the body itself or how the materiality of the body can disrupt the discursive. Instead (and unsurprisingly) social products are believed to be the most concrete things.[9] This highlights the discursive and decidedly not the material nature of social constructionist and post-structuralist approaches. As Butler herself admits, 'I am not a very good materialist. Every time I try to write about the body, the writing ends up being about language.'[10] These attempts reveal the limitations of Elias's *homines aperti*; partly because the social is *always already* inextricably linked to the construction of identity and the experience of the body, even when it was not acknowledged as such. That is to say, even in *homo clausus* accounts where the social is ignored, it still remains part and parcel of the construction of the individual. The social constructionist and post-structuralist take on *homines aperti*, in many ways, replicates the construction of *homo clausus*, but from the other direction: from the outside to the inside. Where *homo clausus* constructs the self within the bounded body, *homines aperti* (de)constructs the body in order to create the self. In both cases the body is rendered into passive materiality, which requires active shaping, management, and control according to social propriety and through social power. *Homo clausus* and *homines aperti* do not do justice to sensory-embodiment because they seek to understand ways of being purely through ways of individualistic, rational knowing; they are only ever theorised epistemologically (based on what we can know or 'is known') even when ontological practices are clearly important to their formulation, because knowing *is* being. This replicates the representationalist spilt between knower and known, observer and observed.

'Ontological theories are about matter; unlike epistemological theories, they cannot "lose" the real—it is their subject matter.'[11] The 'new settlement' [in Latour's (1993) terms] or 'new ontology' that science

studies, posthumanism, and material feminist theories are seeking to
enact is deeply invested in the ontological as known and understood
through the empirical. Linda Alcoff, the American feminist philosopher, describes the ontological as accessible *through* the discursive.[12]
This approach recognises that 'our language structures how we apprehend the ontological but it does not constitute it.'[13] Thus we must be
critically aware of how we arrive at and use discourse in our understanding and construction of knowledge. For example, as Barad states,
'theorizing must be understood as an embodied practice, rather than a
spectator sport of matching linguistic representations to preexsisting
things.'[14] Whereas linguistics and discourse do not even leave room for
sensory-embodiment in their creation of knowledge and understanding,
ontology reveals where we have given cultural representation too much
power.

   While I am in favour of this approach generally, it is even more
useful to describe the approach taken throughout this book using Barad's term: onto-epistemology. Keeping ontology and epistemology separate, even in an approach that recognises how they relate, continues 'a
metaphysics that assumes an inherent difference between human and
nonhuman, subject and object, mind and body, matter and discourse.'[15]
Onto-epistemology acknowledges how ways of being and ways of knowing are not two separate processes, but rather intra-acted. That is,
knowing is only ever possible through being and what and how we know
is contingent on myriad material entanglements. Furthermore, since
ways of knowing and being *matter*, as they are always caught up in
socio-cultural ethics and politics, onto-epistemologies are always also
'ethico-onto-epistemological'.[16] Thus this approach contributes to an
embodied ethics of being; opening possibilities for ethical action. The
concept of 'intra' is also highly useful, as it stresses the inherent un-
sealed nature of being, whereas 'inter' (e.g., in Goffman's [1983] Interaction Order) presupposes an observable and lived separateness between bounded, sealed bodies that exists prior to the social. 'Inter'
follows representational concepts, as explored in the previous chapter,
whereas 'intra' acknowledges the nuance and entanglement of being.
Intra-activity allows us to re-conceptualise modes of being and experientially change ways of understanding, by highlighting the ability of
materiality and ontology to 'feed back' into, indeed to *materialise* discursive epistemologies (by exposing their highly entangled nature).
Intra-activity does not presume that the self exists *within* independently
bound bodies or outside of interdependent yet sensorial individuated,
dis-embodied selves. Rather, it helps us recognise that, as Barad states,
'It is through specific agential intra-actions that the boundaries and
properties of the "components" of phenomena become determinate

and that particular embodied concepts become meaningful.'[17] Similarly, Donna Haraway explains that

> the infolding of *others to one another* is what makes up the knots we call beings or, perhaps better, following Bruno Latour, things. Things are material, specific, non-self-identical, and semiotically active. . . . Never purely themselves, things are compound; they are made up of combinations of other things coordinated to magnify power, to make something happen, to engage the world, to risk fleshly acts of interpretation.[18]

Put simply, bodies are never merely discursive, but rather, 'always already material-discursive'.[19] It is through intra-action that more 'things' we call bodies are condensed into individual selves. *Homo clausus* and *homines aperti* conceptions skip the steps of condensation and begin theorising at the level of the individual, the already assumed self. Furthermore, if we conceptualise the social world in terms of material-discursive practices, as *phenomena through* which agency happens, we can directly access the ways that 'boundaries do not sit still'.[20]

For example, the use of an intra-active onto-epistemology enables a re-imagination of performativity through a posthumanist-materialist lens. This is not merely a set of new terms to describe the happenings of daily life but a revelation of how language is given power *and* how epistemology is always already ontology. As Barad argues:

> Performativity, properly construed, is not an invitation to turn everything (including material bodies) into words; on the contrary, performativity is precisely a contestation of the excessive power granted to language to determine what is real. Hence, in an ironic contrast to the misconception that would equate performativity with a form of linguistic monism that takes language to be the stuff of reality, performativity is actually a contestation of the unexamined habits of mind that grant language and other forms of representation more power in determining our ontologies than they deserve.[21]

Here Barad connects the power granted to discourse to our minds (cognitive processes) which are normally understood as somehow other than our bodies and highlights how our systems of rational understanding restrict our embodiment and thus our possibilities for becoming-other. She says that discourse, as rationally operative, is the epistemological system that *manufactures* the understanding of ontology. Hence, through a representationalist discursive framework, ontology is always already misunderstood at best and utterly mistrusted at worst. This emphasising of Butler's error is quite similar to how Nietzsche pointed out Descartes's oversimplification of nearly the same relationship I refer to in the preceding chapter's conclusion. In Butler's formulation, while performativity may come from outside of the body (that is, because the practices are socially *learned*), ways of being—by being made into mere discourse (i.e., representations)—have no direct recourse to ways of thinking, knowing, or understanding. Here ontology is overtaken completely by epistemology.

Reading Butler's notion of performativity (i.e., 'iterative citationality') through Barad's lens of intra-activity, offers an opportunity to deconstruct and move beyond binary oppositions such as internal/external, self/other, mind/body, active/passive, material/discourse, same/different, ontology/epistemology, subject/object, language/thing, etc. This view does not support Butler's notion of materiality as produced by discourse, but reveals instead how materiality is defined and separated into individual bodies, i.e., made into passive social products. It is vital to understand the process of intra-action in order to contextualise the *feeling*, that is, the *rational embodiment*, of individuality. Intra-action gives relief to how bodies are linguistically differentiated and experientially established as separate and open or closed.

Taken together, these neologisms represent an opportunity for the honouring and exploration of knowing *as* and *through being*. They emphasise an approach to identifying, understanding, and theorising social experience that values, above all else, sensory-embodied knowledge. This is not a return to the reductionist modernism that caused feminist theorists to disown the material body. Rather, the new settlement 'accomplishes what the postmoderns failed to do: a deconstruction of the material/discursive dichotomy that retains both elements without privileging either.'[22] Generally speaking, modern, social constructionist, post-structuralist, and queer approaches either are not concerned with the boundaries of the material self or are interested in fragmenting those boundaries through a privileging of transgression.[23] As Georges Bataille points out: 'There exists no prohibition that cannot be transgressed. Often the transgression is permitted, often it is even prescribed.'[24] Instead, the theoretical method espoused here seeks to put into question and ultimately move beyond the importance of those boundaries entirely by exposing their inherently unfinished, shifting nature. Through the dissolution of *rigid* rationalist bodily boundaries new possibilities for embodied change are revealed and made possible beyond the *homo clausus/homines aperti* binary. In order to experience the de-territorializing effects of threshold experiences, those liberating and generative opportunities for becoming-other, the body must be understood as *corpus infinitum*. This is a move beyond identity politics towards a new ethics and politics of embodiment.

Elias was unsatisfied by the philosophical and sociological tendency to reproduce the mind/body, inside/outside binaries which took individuated, adult identity as the given starting point. His work emphasised process, history, and interconnectedness. Thus social constructionist and post-structuralist approaches that understand *homines aperti* through the fragmentation of body-identity do not fully speak to his project. The shift to a more cohesive approach to body-identity construction requires more than simply giving power to the social from the outside of the body (especially when the materiality of the body is the cost). By honouring Elias's process-oriented sociology, I posit a rethinking of the self as *corpus infinitum*, or boundless, becoming, intra-acted bodies. By starting with bodies, not *selves*, my goal is a material feminist, posthumanist understanding of embodiment that honours sensation and difference in a move towards new perceptual systems of knowing, experiencing, and understanding. One way to engender this sort of dynamic intra-connectedness is through the entanglement sex-gender-sexuality. Instead of conceiving of sex-gender-sexuality as constituted through discursive practices that produce the material body, the entanglement points to the functionality of material-discursive practices.

ENTANGLING MATERIAL-DISCOURSE

In the approaches covered in chapters 1 and 2, the body is decidedly not the experiential starting point in creating, assuming, or reproducing social identity. Rather, it is the socially anticipated and socially constructed *individual*—that is, the rational self who creates dis-embodied experience through a sort of socially taught, self-imposed sensory-embodied filter. This individual is always gendered according to the heterosexual matrix, as it is only through the expectation of heteronormativity that one can observe those who do not follow it. As Colebrook highlights, 'It is because there is a heterosexual matrix that constitutes and delimits subjective possibilities that we could pay attention to those modes of person and enactment that disturbed normative structures.'[25] Heteronormativity is a primary onto-epistemological system of sameness and difference, sensorially embodied through ways of being and knowing.

Working against Butler's blatantly dis-embodied notion of the body as merely the *effect* of discourse,[26] feminist theorists like Luce Irigaray (1996), Elizabeth Grosz (1994, 1995, 2011), Susan Bordo (2000) and Moira Gatens (1996) call for the move away from a degendering politic. Following on from social constructionist and post-structuralist approaches, which are unable to account for the materiality of the body and therefore decide it too must be discourse, a 'degendering politic' seeks to explain away gender so it does not get in the way of theories which *discursively* promote sex/gender fragmentation and fluidity via the phenomenologically 'lived' body. These theories are ultimately unsuccessful in terms of living bodies in daily life because the materiality of personhood remains active and ongoing regardless of sex/gender theory. Furthermore, degendering ignores any possibility of sexual difference *as ontological* difference—where sexed bodies are ontologically different from other sexed bodies from birth—thus allowing for a politics of visibility and equality where, for example, women fight for the right to be equal with men; an impossible project of sameness in *homo clausus* representationalist societies. Degendering ignores the enfleshed reality of every body through a singular axis of sameness. This degendering of the sexually different body through an identity politics of equality is what Irigaray calls the 'new opium of the people'.[27] She explains,

> Some of our prosperous or naïve contemporaries, women and men, would like to wipe out this difference by resorting to monosexuality, to the unisex and to what is called identification: even if I am bodily a man or woman, I can identify with, and so be the other sex. This new opium of the people annihilates the other in the illusion of a reduction to identity, equality and sameness, especially between man and woman, the ultimate anchorage of real alterity. The dream of dissolving material, corporeal or social identity leads to a whole set of delusions, to endless and unresolvable conflicts, to a war of images or reflections.[28]

For Irigaray, those genderqueer or gender-nonconforming people take the denial of the material body even further. Identity politics and the struggle for equality become the project, which is much less radical than Irigaray's project (or mine for that matter) but those gender-nonconforming identities are an unsurprising development in the heteronormative world. They represent a desire to get out from under the singular vision of life and to be different. They speak to *difference.* Yet, queer and trans identities still render the body into passive materiality. This is how I am able to theorise those 'non-normative' experiences with 'normative' ones. Though in their non-normativity, in their speaking to and toward difference, those individuals may be more open to experiences of becoming-other.

What is of particular interest to me in this research is the possibility for becoming, and while queer-, trans-, and heteronormative-identified people all have the potential for becoming-other, as explored in this book, perhaps some queer and trans individuals may have an easier time engaging with those potentials due to their non-normative status. Still both normative and non-normative gender identities participate in a single shared (i.e., patriarchal) world, not one in which 'sexual difference *is* ontological difference, [that is,] the condition for the independent emergence of all other living differences.'[29] For both normative and non-normative identity, difference is representational, not fundamentally generative, because the body is understood as passive materiality. Instead of an emphasis on self-body identity where gender expression is divorced from the materiality of the body, the call for a move away from degendering places embodiment at the core of experience. It recognises that particular styles of embodiment are vital to how people actually experience their body-selves and the desire to flatten those experiences through theoretical discourse does little to change how people are embodied, that is, how people live each day.

While many feminist and queer theories have sought to divorce sex (i.e., 'biology') from gender (i.e., the cultural expression of the 'sex'), I do not find this distinction useful or even possible when understood through material-discourse. It is only possible linguistically, *discursively* because we learn how to be and have bodies based on heteronormative sex/gender. As Butler explains, 'Indeed, if gender is the cultural significance that the sexed body assumes, and if that significance is codetermined through various acts and their cultural perception, then it would appear that from within the terms of culture it is not possible to know sex as distinct from gender.'[30] What's more, sex and gender are absolutely condensed *experientially*, empirically, since one cannot get out from 'within' their culture. For example, toileting practices are foundational to socio-cultural understandings of sex-gender. Regarding Lacan's story of the 'laws of urinary segregation',[31] MacCannell explains, 'same sex bathrooms are social institutions which further the metaphorical work of hiding gender/genital difference. The genitals themselves are forever hidden within metaphor, and metaphor, as a "cultural worker", continually converts difference into the Same.'[32] Genitals (i.e. materiality) are usually hidden by clothing, averted gaze, or physical barriers (i.e., more materiality) which *symbolically produce* gender (i.e., discourse); this condensation of materiality into discourse via epistemological expectations and anticipation—we assume that we know (based on sensory experiences that are construed as rational) how people will act, what they will look like, and thus how they are sexed according to practices (i.e., ways of being) that are collectively taken as gender—is how matter is understood as discourse. It creates sameness of bodies where there is difference (because all bodies are always already different).

This is similar to how *homo clausus* identity is *naturalised* according to its many binaries. Therefore, as long as one's expression of social identity conforms to the gender listed on the toilet room door, the material makeup of one's genitals is of no real consequence (though it is rare that people actually transgress these social codes without good reason, like the many trans and genderqueer people who have not undergone sexual reassignment surgery and use these spaces daily, often fearful of the materiality of their bodies being seen as not matching the gendered discourse they 'give off'). What is vital here is recognising where materiality shows up in daily social practices and self-body experiences, but is condensed into the rational and discursive. The above example, regarding same-sex toilet spaces, is one way material-discourse works by giving relief to some embodied practices which help maintain *homo clausus* individuality (via rational classification of the sameness/difference of bodies), and also an example of where we can begin to cultivate sensory-embodied awareness as a living source of knowledge and understanding. I am not suggesting that we all need to start looking at one another's genitals in public toilets, but rather noticing when and where our sensory-embodied practices are rendered into flat discursive understandings and disconnected through socially instilled feelings like fear, anxiety, shame, and embarrassment. Materiality is always linked with discourse because our *lives are material*. Bodies being hidden, eyes being averted, and standing behind an erected barrier are all sensory-embodied practices, *not* discursive practices. These material happenings may give way to discourse, to epistemology, but they are fundamentally ways of being.

Thus the turn away from materiality in queer, postmodern, and feminist theories has enabled the *linguistic* distinction of sex, gender, and sexuality. These three 'identity categories' are vital to one's 'basic identity' in the *homo clausus* and *homines aperti* conceptualisations. Moira Gatens, who understands the body *not* as passive materiality onto which gender is discursively sculpted, points out that the body is not neutral to begin with. She argues: 'If one wants to understand sex and gender or, put another way, a person's biology and the social and personal significance of that biology',[33] then what we need is to understand the body as *lived*.[34] Therefore the post-structuralist and social constructionist separation of sex (biology), gender (social expression of sex), and sexuality (what/who/how 'one' desires) is another example of granting discourse too much power over the experiential. Furthermore, these distinctions ignore that linguistics, words, thoughts, and discourse are all material process too. They are not other than matter, but rather materially entangled and entangling phenomena. Thus rendering sex, gender, and sexuality distinct classifications can work epistemologically, but it does not work ontologically—instead of reducing these elements to material-ity *or* sociality, it is more useful to understand them onto-epistemolog-ically. Otherwise, they will remain *either* in the realm of discourse *or* the material and prevent any new ways of being, knowing, and under-standing.

The atomisation of sex, gender, and sexuality into three distinct cate-gories is what enables the discursive phenomenon of fragmented iden-tity. When the body is understood as entirely de-territorialized, until social discourse is used to re-territorialize it, one's identity is fragment-ed from one's materiality. This requires a rational separation that flat-tens embodiment into mere biology and produces a discontinuity of experience in the attempt to create stable sameness. This gets us no-where in terms of sensory-embodied experience, since we cannot sim-ply turn off or disengage parts of ourselves, even if we believe this is conceptually possible; the body persists. Likewise, the process of senso-rial individuation, which seeks to separate and control the 'senses', does not allow us to willingly disengage from parts of our bodies entirely. Therefore, these categories must be brought into conversation, pulled from their neat classifications outside of the body and immersed into sensory-embodiment. Sex, gender, and sexuality are always already con-densed according to heteronormative (and homonormative)[35] *homo clausus* assumptions of the body via material-discursive practices.

To echo Irigaray, we live our identities through a sexed body. Em-bodiment is primary, and however we may live our gender or express our desire, it all happens with and through our flesh. While sex, gender, and sexuality, as three distinct elements, may create the *impression* of variable identity through dis-embodied *homines aperti* fragmentation, it is only variable when compared against the rational heteronormative matrix. In order to begin shifting conceptions of social embodiment from the closed/open binary to knowing through doing and being, it is important to recognise these elements are inherently entangled. Bodies are messy, dynamic, and unfinished, and it is time that conceptions of subjectivity acknowledged this. This is not a call for a return to natura-listic, reductionistic, or deterministic approaches to the material body. Instead, sex-gender-sexuality, in whatever combination, is a move to-wards the *living* body and sensory-embodied knowledge. Sex-gender-sexuality, as an entanglement, points to material-discursive bodies in action, allowing for difference, exploration, and experimentation be-yond terminology and categorisation. This entanglement happens in time and space. It is material and discursive, ontological and epistemo-logical, differential and becoming, as opposed to separate categories that are rationally constructed onto the material body, or naturally oc-curring from within.

While sex, gender, and sexuality are often most immediately associated with the domain of the 'inner' life and thus one's private lifestyle, they are absolutely vital to the experiencing of everyday public life because they are assumed to line up according to the discourse of/discursive heterosexual matrix. For example, as Siebers, when speaking about differently abled bodies engaged in sexual acts, explains, 'Sex may seem a private activity, but it is wholly public insofar as it is subject to social prejudices and ideologies and takes place in a built environment designed according to public and ideal conceptions of the human body.'[36] Even the most seemingly private acts are subject to social standards of heteronormative sex-gender-sexuality. Therefore, it should come as no surprise that the most mundane aspects of daily life (e.g., toileting practices) are too. This is why the styling of the body *through* sex-gender-sexuality is so crucial. As Peggy Phelan explains, 'Identity emerges in the failure of the body to express *being* fully and the failure of the signifier to convey meaning, exactly.'[37] Therefore the entanglement sex-gender-sexuality is more useful to an onto-epistemological understanding of the embodied self, as it speaks to how bodily experience is always already caught up in myriad ways of being and understanding that cannot be cleanly parsed out into three separate categories. It points to the materiality that is caught up in clandestine social-rational process that condenses to categorise and name bodies, practices, and desires. Sex, gender, and sexuality are not preformed social variables, but rather ongoing, open-ended sensory-embodied processes that are part and parcel of the materiality of the body. They are variable and changing depending on how one is embodied, not stable features of identity. To keep them separate denies the entangled, differential, and becoming nature of living.

To flesh this out a bit more and from a different approach, I will now consider the reproduction of sex-gender-sexuality as something which is clearly entangled but is rationally understood and thus taught/learned as distinct categories. Like Elias's annoyance at the tendency of Western philosophy and sociology to take the adult *homo clausus* subject as given, a criticism associated with his espousal of the need for empirical awareness of the processes of becoming and being an individual in society, Lee Edelman is fed up with the heteronormative project of 'future children'. This project shores up social individualism and the *homo clausus* self in the form of families who reproduce this individualism in and through their offspring. As the process of child-rearing is highly social, it is extremely difficult to raise children outside of or in opposition to it. Heteronormativity is reliant upon two distinct sex-genders to further the mission of the linear progression of the species, in the form of families that occur as distinct units consisting of a 'mother' and a 'father'. According to Edelman, 'Fighting for the future of the children' is arguably the ultimate heteronormative political and moral project—as it is necessary for the reproduction of capitalist individualism—and there is an immense amount of social inculcation into this project that occurs in childhood. Edelman works, in his polemic *No Future: Queer Theory and the Death Drive*, to unpack it. He argues, '*queerness* names the side of those *not* "fighting for the children," the side outside the consensus by which all politics confirms the absolute value of reproductive futurism.'[38] This idea of the family and those who should make up families is a fantasy that provides a false sense of stability (sameness) to gendered social identities and related bodily ways of being according to sex-gender-sexuality.[39] As I work to show in this book, heteronormative practices imbue our ways of life generally, even for those who are not heterosexual. This is largely due to how we construct gender identity and sensory-embodiment based on heteronormative ideals, which entangle sexual ways of being with the sexed and gendered body. In other words, 'one way in which [the] system of compulsory heterosexuality is reproduced and concealed is through the cultivation of bodies into discrete sexes with "natural" appearances and "natural" heterosexual dispositions.'[40] Gay or straight, male or female, this naturalisation is primary for individual, rational *homo clausus* identity, because sexuality and gender are reliant on the materiality of the body. Instead of rationally separating sex, gender, and sexuality to combat the already overly rational *homo clausus* sex-gendering, we need to return to the materiality that is suppressed in identity construction. In order to do so we need to learn to recognise patterns of difference (without categorising and pathologising them), which are always already present for all bodies every day. The economic, ethical, and moral project for the 'future for children' is concerned with sameness and seeks to demonise difference according to heteronormative ways of being. This dynamic is even at work in public toilet spaces.

Children and their 'innocence' are of a specific concern when it comes to public toilets and the need for bodies to replicate heteronormative sex-gender-sexuality. It is through their invocation in public toilets that queer and trans bodies are made into something to be suspect and scared of, and often those individuals are labelled as 'perverts' or 'paedophiles' simply because of the fear surrounding non-normative sex-gender-sexuality. This was reflected in my interviews with both queer women and trans men, who dislike when children are in public toilets because they tend to be curious and not fully inculcated into *homo clausus* norms and thus are thought to be able to spot difference, a curiosity that could result in abuse from angry, fearful parents. Part of this concern is due to how sexuality is always entangled with sex-gender because it is never *just* about how and with whom one has sex, but instead the entire implied or assumed lifestyle, ethics, politics, and morality that is seemingly expressed through one's embodied self—that is, one's identity. This is an example of material-discourse at work. In public toilets, where sex-gender is condensed, sexuality that is not normative is expected to be able to be accessed through appearance and bodily behaviours, through the visual, so that even those who do not 'identify' with heteronormativity experience pressure to embody the socially 'proper' behaviours so as not to appear suspect and threatening. This is because according to *homo clausus* norms there is a right and a wrong way to be embodied, hence the rational expectation to be able to access one's sex-gender-sexuality through a material-discursive reading of one's body—a reading which is understood not as fleeting or momentary, but rather an image of stability that gives access to one's entire way of being according to normative social structures.

This bodily reading of one's presumed identity is a representationalist practice that seeks to create sameness amongst identity 'types' and can only account for difference in oppositional terms. Like Grosz, my interest in this book is 'in addressing how difference problematises rather than undergirds identity'[41]—that is, a difference which is the ontological condition for the social, a material, sensory-embodied difference that is before and beyond 'identity', 'subjectivity', or 'consciousness'. This 'is an understanding of difference as the generative force of the world, the force that enacts materiality (and not just its representation), the movement of difference that marks the very energies of existence before and beyond any lived or imputed identity.'[42] Put simply, and to reiterate, difference is the condition for being in the world and differential ways of being do not stop appearing no matter how intensely we normatively and continually (re-)territorialize ourselves (via identity, social propriety) against them. Difference is before and beyond any categorisation of sameness from which superficial, representational difference can be located, and *corpus infinitum* is a way beneath identity and into difference through sensory-embodiment. Sensory-embodiment, when given the opportunity, can open those self-variations and experiences we learn to ignore, hide, or neglect according to social practices of teleological sameness that bring us beyond the self, or 'make us *more than we are*'.[43] In *Difference and Repetition,*

> Deleuze outlines how the concept of difference is aligned, repressed, and evaded in the history of Western thought, but also the ways in which nevertheless a monstrous, impossible, unconstrained difference is implicated in all concepts of identity, resemblance, and oppo-

sition by which difference is commonly understood and to which it is
usually reduced.[44]

This point is also why sex-gender-sexuality is an important way into the
understanding and valuing of difference. The triad removes the possibility for definition and stability based upon and between identity
categories and, rather, makes room for ongoing difference as inherent
to sensory-embodiment. No longer are we subjects with specific sexes,
genders, and sexualities, attempting to cobble together a stable, self-
same experience from discontinuity, but instead are open, becoming
entities with various features of materiality and opportunities for ways
of being that enable a smoother, more cohesive daily experience. Rather than allowing difference to be 'subjected to representation', *corpus
infinitum* is an attempt to move beyond and get before 'identity, analo-
gy, opposition, and resemblance' which transform difference 'from an
active principle to a passive residue'[45]—that is, to give difference its
place as generative force. This, I believe, is possible when identity is not
based on the categorisation and pathologisation of ways of being, which
require individuals to ignore, exclude, or refrain from the curious, imaginative, and creative opportunities that do not perfectly align with one's
'identity'. This social restriction through identity creates fragmentation
and disconnection while trying to produce sameness.

Thus, by acknowledging sex-gender-sexuality as experientially entangled according to social norms that can be accessed by others through
the visible, we can begin to locate where discursive sameness overtakes
material difference. Phelan echoing Lacan reminds us that 'visibility is a
trap. . . . It summons surveillance and the law; it provokes voyeurism,
fetishism, the colonialist/imperial appetite for possession.'[46] Social identity is reliant upon the specific control, management, and styling of the
body, and those stylings are always caught up in heteronormativity (and
in the project of the family, even when the parents aren't heteronormative, because heteronormativity is the basis for both the family and sex-
gender-sexuality). The categories of sex, gender, and sexuality, while
perhaps useful for linguistic distinctions and the construction of a
'stable' 'identity', which is not consistently 'reflective' of sensory-embodied practices (because reflective categories ignore subtle patterns of
difference), are inseparable in the daily experience. For that reason sex-
gender-sexuality as an entangled onto-epistemology is more accurate
for an empirical study of sensory-embodiment and a vital opening into
daily patterns of difference.

MATERIALISING AND POSTHUMANISING ELIAS

The awareness, arguably developed out of Butler's brand of performativity, that 'self-identity needs to be continually reproduced and reassured precisely because it fails to secure belief' is a starting point for understanding how onto-epistemology happens through repetitive material-discursive practices.[47] Barad's grounded performativity, which 'can only be understood as an explicit challenge to Butler's concept', offers an opportunity to recognise where it makes sense to actively intervene in the reproduction of heteronormative *homo clausus* body-identity.[48] Reading Elias through material feminist and posthumanist theories enables a perspective of how bodies are transformed into and experienced as individual selves. "'Humans" are neither pure cause nor pure effect, but part of the world in its open-ended becoming.'[49] Reading Elias through Barad affords a new opportunity to not only *observe* and *describe* human society as *homines aperti*—consisting of individuals who are 'interdependent people in the singular' and society as 'interdependent people in the plural'[50]—but to account for the intimate, entangled experience of living and being by bringing this epistemological view into the ontological, through the intra-active *corpus infinitum*.

Homo clausus, the typical individual identity formation and experience in the English-American context, is based upon 'a cultivation of inwardness' and an honouring of the rational self, who *has* agency.[51] But *homo clausus* is much more than merely an 'image' of the self that took hold over time as Smith claims;[52] *it is an entire ontology*. In fact it is an onto-epistemology, because knowing and doing are always entangled, and intra-acted. *Homo clausus* is not just how people construct their body-identity, their 'self', but it is how people *experience, know*, and *understand* their bodily selves. Without the strict and rational shaping of the experiential body, the self would not be able to condense into an 'image' or 'stable' identity. Therefore, I view this conjoined theoretical opportunity as a potential for sensory-embodied identity to not only be recognised and empirically theorised as open, fluid, and interconnected (e.g., *homines aperti* or queer), but that a profound shift in the experiential, the living self, is possible when we place sensory-embodiment at the core of our understanding. This is a re-focusing of what we take to be real and powerful and a re-conceiving of those relationships. By placing *active* materiality at the centre of power there isn't a denial of rationality, the mind, epistemology or theory, but rather, a new more cohesive, expansive, and entangled approach to being beyond dialectics. Both *homo clausus* and *homines aperti* understandings of selfhood and the social are inherently limiting because they attempt to create stability and legibility through material-discursive practices that deny inherently open-ended, ongoing difference. As Grosz explains,

> Oppression is made up of a myriad of acts, large and small, individual and collective, private and public: *patriarchy, racism, classism,* and *ethnocentrism* are all various names we give to characterize a pattern among these acts, or to lend them a discernable form. I am not suggesting that patriarchy or racism don't exist or have mutually inducing effects on all individuals. I am simply suggesting that they are *not* structures, *not* systems, but immanent *patterns,* models we impose on this plethora of acts to create some order.[53]

Similarly, *corpus infinitum* seeks to get before and beyond the systems, the structures, to the patterns that condense into individual identity and keep people from the becoming nature of sensory-embodiment. This is a potential to shift from individual identity to immanent patterns of being and from social structures to greater personal empowerment.

By first locating (through empirical research) fissures in the monadic self-image (via socially experienced identity) and then exploring how those fissures are felt in relation to the individual body, I believe this shift is possible. Put simply, instead of a social identity based on disembodiment for the sake of social connectivity, by way of maintaining classificatory systems, this is a move towards sensory-embodiment that is not swiftly weakened through fear, anxiety, shame, and embarrassment. Through sensory-embodied cohesion, the loosening of internal and the external borders, and a devaluing of boundaries created by discourse, the power of FASE to regulate individuality can be radically diminished. This is extremely important not merely theoretically, but empirically and onto-epistemologically. Sensory-embodied relationships, as reliant on the notion of fixed or fragmented boundaries of the individual body (*homo clausus* or *homini aperti*) exclude an entire range of embodied possibilities, let alone access to 'important dimensions of the workings of power'.[54] By giving attention to enfleshed potential I aim to empirically explore how the dominant system of individual identity creates dis-embodiment and thus severely limits us in our social connections. In doing so, I seek to extend Elias's project beyond the binaries, biases, and material losses inherent to *homo clausus* and *homines aperti* and offer an understanding of the bodily self as *living*. By taking sensory-embodiment as the source of experience, knowledge, and understanding—not the rational mind—new socially experiential possibilities can be forged.

CONCLUSION AND CONNECTIONS

Posthumanisms, material feminisms, and Eliasian sociology stress the need for empirical research. The empirical research within this book aims to build a cohesive English-American understanding of sensory-embodiment through Eliasian understandings of identity and society. This work is situated in relation to Mennell's Eliasian study of America, wherein he shows that

> the broad trend of the development of manners and habitus in the USA was very similar to that which Elias observed between the late Middle Ages and the nineteenth century in Western Europe, and later studies (notably by Cas Wouters) suggest that in the twentieth century the pattern was one of further convergence. . . . On both sides of the Atlantic, changes in social standards were driven by similar, though not identical, processes . . . [and] through a growing social constraint to self-constraint, the same advance in the thresholds of shame and repugnance is evident in American [*sic*] as in Europe.[55]

In my research, I have focused on the experience of sensory-embodied living in gender-segregated English-American public toilet spaces. Since these societies have developed and construct identity and self-control in very similar ways—that is, based upon similar ideals of gender and morality that is expressed, for example, through an engagement with patriarchal capitalism, including media and consumerism—it is possible to theorise them together. This offers a better opportunity to give relief to the naturalised taken-for-grantedness inherent in *homo clausus* identity construction, as it allows access to more types of identity. Public toilets, simply put, bring together individuality, social identity, the biological body, sensory individuation, dis-embodiment, sex-gender-sexuality and experiences of fear, anxiety, shame, and embarrassment on a daily basis. They are spaces where monadic *homo clausus* thrives through *homines aperti* fragmentation of the self as separate from the 'disgusting' or 'abject' material body.

Public toilet spaces also offer a vital link to the body as living *before adulthood*. As Elias stresses, 'The historicity of each individual, the phenomenon of growing up to adulthood, is the key to an understanding of what "society" is.'[56] Toileting practices are (expected to be) taught and learned once and for all. Once we learn them in the initial years of life, they become tamed by taboo and we are not to speak of them again. This is another example of material-discourse in action. Toilet training for parents of toddlers is actively and enthusiastically discussed for a very brief (and often frantic) time, then the discussion is abruptly quashed when it is expected that all children have rationally learned what they 'needed to learn' in order to control their bodies accordingly. These are some of the earliest and deepest experiences of dis-embodiment, and they are some of the most long-lasting sensory-embodied practices we sustain in our performances of sex-gender-sexuality and embodied identity generally.

By taking the living adult toileting-body seriously it is my goal to better understand social identity as necessarily tied to sensorial individuation and practices of dis-embodiment from our earliest social relationship with bodies. In doing so I elucidate how individual social identity maintains socially legible bodies and continues the valuation of social connection based upon reductionistic dis-embodiment. Through prying open individual experiences of social identity (as required for social connectivity) and putting them into conversation, I point to strategies people have adopted both by choice and out of necessity, that further a richer embodied experience, which I describe as bodily becoming. These experiences explore an embodied shift from *homo clausus* to *homini aperti*—from closed self to open interconnected self—and beyond; from dis-embodiment reliant on consistent patterns of sensory individuation to open ongoing sensory-embodiment as boundless becoming, as intra-active social happening. Ultimately, it is a shift that explores the potentials of bodily becoming above mere social connection, by directly taking on fear, anxiety, shame, and embarrassment and exploring personal and non-conventional forms of embodied knowledge as the very thresholds for becoming-other. Through this exploration I hope to reach an understanding of how we can become more than merely the sum of our parts.

The posthumanist-materialist formulation of sensory-body-identity is about taking a diffractive and onto-epistemological approach that aims to recognise, understand, and disrupt limiting material-discursive practices inherent to *homo clausus* or *homines aperti* ways of being, knowing, and experiencing. This approach frees the mind from the imagined boundaries of the skull and allows for bodily awareness, for *knowing through the consciously aware body*. By taking a posthumantist and material feminist approach to my research I aim to understand body-identity as *living*, not rationally *maintained*. Sensory-embodiment is about knowing through *being* not merely *doing* based upon knowing. Toileting practices and public toilet spaces offer an opportunity to understand where and when *homo clausus* identity breaks down and therefore how the shift to *corpus infinitum* can happen.

NOTES

1. Karen Barad, *Meeting the Universe Halfway: Quantum Physics and the Entanglement of Matter and Meaning* (Durham, NC: Duke University Press Books, 2007), 137.

2. Kristin Ross, *The Emergence of Social Space: Rimbaud and the Paris Commune* (London: Verso, 2008).

3. Donna Haraway, 'The Promises of Monsters: A Regenerative Politics for Inappropriate/d Others', in *Cultural Studies*, ed. Lawrence Grossberg, et al. (New York: Routledge, 1992), 300.

4. Barad, *Meeting the Universe*, 29.

5. Barad, *Meeting the Universe*, 29.

6. Haraway, 'The Promises of Monsters', 318, my emphasis.

7. Karen Barad, 'Posthumanist Performativity', in *Material Feminisms*, ed. Stacey Alaimo and Susan Hekman (Bloomington: Indiana University Press, 2008), 120.

8. Barad, *Meeting the Universe*, 64.

9. Alexandra Maryanski and Jonathan H. Turner, *The Social Cage: Human Nature and the Evolution of Society* (Stanford, CA: Stanford University Press, 1992), 105.

10. Judith Butler, *Undoing Gender* (New York: Routledge, 2004), 198.

11. Susan Hekman, 'Constructing the Ballast: An Ontology for Feminism', in *Material Feminisms*, ed. Stacey Alaimo and Susan Hekman (Bloomington: Indiana University Press, 2008), 98.

12. Linda Alcoff, 'The Problem of Speaking for Others', in *Who Can Speak: Authority and Critical Identity*, ed. Judith Roof and Robyn Wiegman (Urbana: University of Illinois Press, 1995).

13. Hekman, 'Constructing the Ballast', 98.

14. Barad, *Meeting the Universe*, 54.

15. Barad, 'Posthumanist Performativity', 147.

16. Barad, *Meeting the Universe*, 185.

17. Barad, 'Posthumanist Performativity', 133.

18. Donna Haraway, (2008, p. 250)

19. Barad, 'Posthumanist Performativity', 141.

20. Barad, 'Posthumanist Performativity', 135.

21. Barad, 'Posthumanist Performativity', 121.

22. Stacey Alaimo and Susan Hekman, *Material Feminisms* (Bloomington: Indiana University Press, 2008), 6.

23. Claire Colebrook, 'How Queer Can You Go? Theory, Normality and Normativity', in *Queering the Non/Human*, ed. Noreen Giffney and Myra J. Hird (Hampshire, UK: Ashgate, 2008), 21.

24. Georges Bataille, *Erotism: Death and Sensuality* (San Francisco: City Lights Books, 1986), 63.

25. Colebrook, 'How Queer Can You Go?' 21.

26. Judith Butler, *Bodies That Matter: On the Discursive Limits of 'Sex'* (New York: Routledge, 1993), 30.

27. Luce Irigaray, *I Love to You: Sketch of a Possible Felicity in History* (London: Routledge, 1996), 60–62.

28. Irigaray, *I Love to You*, 60–62.

29. Elizabeth Grosz, *Becoming Undone: Darwinian Reflections on Life, Politics, and Art* (Durham, NC: Duke University Press, 2011), 105.

30. Judith Butler, 'Performative Acts and Gender Constitution: An Essay in Phenomenology and Feminist Theory', *Theatre Journal*, 40, no. 4 (1988): 524.

31. Jacques Lacan, *Ecrits: A Selection*, trans. Alan Sheridan (London: Routledge, 1989), 167.

32. Juliet MacCannell quoted in Peggy Phelan, *Unmarked: The Politics of Performance* (London: Routledge, 1993), 151.

33. Moira Gatens, *Imaginary Bodies: Ethics, Power and Corporeality* (New York: Routledge, 1996), 11.

34. Hekman, 'Constructing the Ballast', 107.

35. Hetero- and homo-normativity both rely on two distinct and stable sexes, genders, and sexualities. They encompass more than *just* sexuality because they implicate how the body is used generally in everyday life according to my understanding of sex-gender-sexuality as fundamental to social identity and bodily experience.

36. Tobin Siebers, 'Disability Experience on Trial', in *Material Feminisms*, ed. Stacey Alaimo and Susan Hekman (Bloomington: Indiana University Press, 2008), 298.

37. Phelan, *Unmarked*, 7, my emphasis.

38. Lee Edelman, *No Future: Queer Theory and the Death Drive* (Durham, NC: Duke University Press Books, 2004), 3, original emphasis.

39. Edelman, *No Future*, 7.

40. Butler, 'Performative Acts and Gender Constitution', 524.

41. Grosz, *Becoming Undone*, 91.

42. Grosz, *Becoming Undone*, 91.

43. Grosz, *Becoming Undone*, 91, original emphasis.

44. Grosz, *Becoming Undone*, 92.

45. Grosz, *Becoming Undone*, 93.

46. Phelan, *Unmarked*, 4.

47. Phelan, *Unmarked*, 4.

48. Hekman, 'Constructing the Ballast', 104.

49. Barad, 'Posthumanist Performativity', 139.

50. Norbert Elias, *What Is Sociology?* (New York: Columbia University Press, 1978), 125.

51. Norbert Elias, *The Civilizing Process: Sociogenetic and Psychogenetic Investigations* (Oxford: Blackwell, 2000), 512.

52. Dennis Smith, 'The Civilizing Process and the History of Sexuality: Comparing Norbert Elias and Michel Foucault', *Theory and Society*, 28, no. 1 (1999): 85.

53. Grosz, *Becoming Undone*, 97, original emphasis.

54. Barad, 'Posthumanist Performativity', 144.

55. Stephen Mennell, *The American Civilizing Process* (Cambridge: Polity Press, 2007), 295.

56. Norbert Elias, *The Society of Individuals* (London: Continuum, 1991), 25.

II

Individuating the Communal

4

THE HISTORY OF WESTERN PUBLIC TOILETS SINCE THE FIFTEENTH CENTURY

There are several histories that can be written which take as their focal point the locus of public toilet spaces in the Western world. Public toilet spaces impact, historically and presently, upon constructions and experiences of individuality, race, class, gender, age, ability, mobility, morality, sexuality, health, sanitation, technology, and urbanity. They are essential spaces for Western civilisation. It is more than a little surprising, however, that if we consider the social and technological advancement that the word 'civilisation' implies, these toilet spaces as we know them today (i.e., gender segregated with individual, enclosed flush toilets) stopped developing almost as soon as they came into being. As Lambton points out, the toilet 'was invented some 100 years ago [and] since the 1880's has changed neither its workings nor its basic shape.'[1] This lack of advancement is not global; high-tech Japanese toilet and fixture manufactures have spent millions trying to break into the Western market, but have found that Westerners just 'don't care' about advancements in toileting styles and technology.[2]

The public toilets familiar to those in the United Kingdom and North America, which were developed from early western European models, namely French and English, have not changed much since the late 1800s, and neither arguably have our bodily dispositions while we use them.[3] As Alexander Kira, in *The Bathroom*, his unique, important, and severely overlooked[4] architectural study of the way modern people physically use bathing and toileting facilities, reminds us: '[T]echnology is to a very large degree a variable that can be speeded or slowed according to the social and cultural demands of an era. While we can create new technologies to satisfy our demands, we can also ignore particular technologies and allow them to lie idle for years.'[5] Public toilets took several hundred years to come into fashion and, in their earliest forms, were only considered necessary by the ruling elites when continued urbanisation was threatened by the increasing amount of human waste being thrown out of windows, piling up in the streets, and contaminating the bodies of water necessary for sustaining life. As people settled together, creating more densely populated areas, there needed to be some organisation of the human waste that was threatening to submerge the city, hinder mobility, and threaten the health of the population. The privatising process around which this organisation developed began in the home with chamber pots and cesspits and eventually moved out to the public realm and onto the street with specific spaces designated and then built for excretion. This eventually led to the development of flush toilets and sophisticated plumbing and sewerage systems we are familiar with today, systems that whisk our bodily waste away from us in an instant.

Kira's point becomes exceedingly clear when we consider that the flush toilet, a technology Westerners not only rely on but *always* expect in daily life, was first designed and gifted to royalty at the end of the sixteenth century,[6] but did not gain popularity until the Victorian obsession with health and hygiene took hold roughly 200 years later.[7] It was only through the development of modesty, guilt, and privacy that the technology became vital, a point I develop later when considering the cultural shift from *gemeinschaft* (community) to *gesellschaft* (society). The acceptance and development of public toileting was a 'civilising process' in its truest sense; implicating not only the development of technology and new living standards, but also an intense overhauling of people's understanding of and attitudes towards their own bodies and the bodies around them. This process arguably created a new, fragmented, and individuated form of embodiment, one that approximates *homo clausus* ways of being.

THE HISTORY OF PUBLIC TOILETS FROM THE COMMUNAL TO THE PUBLIC

Clara Greed, one of the few toilet space scholars, claims that 'there have always been public toilets.'[8] This rather ahistorical statement, rife with modern mentality, misses the nuance of what it means for a space, particularly a space built for the body, to be 'public'. (We could call the street a 'public toilet', but it will still function primarily as a street.) In order for a space to be labelled and experienced as 'public' there first needs to be a sense of 'publicness' (and a parallel sense of 'privateness'). Particularly when considering toilet spaces, there needs to be bodily modesty and the sense of personal *privacy*. Without this the space is merely communal. It is more useful to say without euphemism that there have always been public acts of urination and defecation. In the following section I will first broadly consider the public and private, personal and communal aspects of toileting and, second, begin focusing the discussion more specifically on three spatial-historical milestones of public toileting.

Historically, the spaces where excretory acts took place (without modern shame and embarrassment) ranged from the communal to the personal to the common. In the medieval era, people would urinate or defecate in all manner of different spaces; whether these be in a field or street (still acceptable in many places for men and children), or while sitting at a table, in the corner of a room, or on the stairs in one's home. Medieval cities did sometimes possess latrines which sat anywhere from eight to twenty-eight users at one time. In contrast, beginning in the 1700s, some women had portable glass, leather, ceramic, or wood urinals which enabled them to urinate anywhere, at any moment, without staining their heavy, layered clothing.[9] Those who didn't have these urinals still urinated in public, but simply risked wetting their clothes.

In all of these cases there was a lack of distinction between 'public' and 'private'. There were many overlapping options for excretion in public life throughout the centuries and of those mentioned above, the one that most closely resembles the public toilets of today (medieval latrines) was favoured for a purpose that would, in the twenty-first century, be considered (according to the rules of the *homo clausus* triadic intra-action order I put forth in chapter 5) utterly undesirable: they were spaces to socialise *during* excretion.[10] Thus at a time before modern guilt, modesty, shame, embarrassment, and privacy, latrines were the 'public toilet' of choice in the fifteenth through eighteenth centuries and were probably utilised *precisely because they were communal*. As Kira notes, 'Not only was defecation simply not always private; it was also often an activity to socialize over' and not only for the lower classes who had more relaxed codes of social propriety.[11] As Wright explains, 'Kings, princes and even generals treated it [the latrine, basin, or whatever apparatus favoured at the time] as a throne at which audiences could be granted.'[12] From commoners to royalty, toileting was often enjoyed as an opportunity to be with others. Further, it has been suggested that 'shared toilet facilities engendered a sense of community among users that would be the envy of modern town planners trying to regenerate the inner city.'[13]

This communal aspect is manifestly absent from the 'private acts' we accomplish in the public toilets of today; communality as explored in the following chapter is not only absent but anathema to contemporary public toileting. Socialisation into the 'proper' use of these spaces and our bodies within them has required communities to become individuals and for individuals to become quiet masters of their bodies, learning strict control and management in accordance with social propriety or otherwise risk personal degradation and social embarrassment. This is exemplified in the toilet training process Western parents put their children through today, with children having 'in the space of just a few years to attain the advanced level of shame and revulsion that has developed over many centuries.'[14] This is a point I develop generally throughout this chapter and deal with explicitly in the final section on embodiment and individuality.

In order to better understand how we sustained this massive shift from toileting spaces that were enjoyed because of their communality, openness, and sociality to the closed and intensely managed spaces of modern public toilets,[15] I will focus on three spatial-historical milestones and their corresponding attitudes and developments in sociobodily practices. These milestones are anchored to specific design or architectural developments that highlight how the (*homo clausus*) self-body disposition we tend to adopt towards bodily functions today was developed and directly entangled with social discourse and material everyday engagement throughout history. It is worth briefly noting that in terms of types of embodiment as well as toileting technology, these changes were not automatically welcomed, supported, or embodied and certainly not adopted without conflict. I have chosen just three for ease of explanation but also because there weren't that many to choose from. As mentioned previously, very little has changed about these spaces since they came into being. Thus the three developments I have chosen show how slowly these changes developed and how quickly cessation of development set in. To accomplish this task I rely largely on Elias's *The Civilizing Process*. His analysis is highly relevant to my study overall and particularly useful in this chapter because of his development of the European habitus—that is, his keen tracing of how socially based emotional modelling and physical behaviour (e.g. manners, violence, repulsion) was shaped, made relevant, and gradually expected in all individuals, a large portion of which we continue to adopt as 'second nature' today. While his work doesn't focus explicitly on the body, much of his observations, particularly in the first volume of *The Civilizing Process*, focus explicitly on bodily functions and the development of shame and embarrassment connected to those functions. Accordingly, my history focuses on the body in public toilets (i.e., the development of the modern, *homo clausus*, toileting body) and begins in the Tudor era: Starting with the 'great house of easement,' then moving to the first gender-segregated toileting spaces at an elite French ball in 1739, and finally arriving at the first permanent, public-private, gender-segregated toilets that resulted from the Great Exhibition and the Victorian obsession with hygiene and sanitation. I then consider the intrinsically linked spatial-historical and socio-cultural shift to *gesellschaft* via individuation, and the resulting implications for embodiment (which I more directly flesh out in the following chapter). My suggestion throughout this account is that the history of the modern public toilet can be interpreted as one of ever-increasing separation, suppression, and hetero-sex-gendering in the name of moral propriety, safety, health, and hygiene.

THE FIRST SPATIAL-HISTORICAL MILESTONE

The latrine, a nascent multi-user public toilet, is the first spatial-historical milestone in toileting technology. It was the communal option of the fifteenth through early eighteenth centuries and was simply a long bench with several holes over which people sat. Latrines were usually built over a body of water (e.g., the edges of bridges over the River Thames in London) into which expelled waste would directly drop. Latrines were viewed as spaces in which group identity could be encouraged and celebrated (or read differently, where group identity could be shaped and then observed), which is why they are utilised, even in the twenty-first century, by 'highly structured and authoritarian' institutions (e.g., the military) who are seeking 'to minimise individual identity, so as to foster or force a strong group identity.'[16] A royal version of the latrine, used by courtiers and servants at Hampton Court in London, called the 'great house of easement', was built in the early sixteenth century during Henry VIII's massive expansion of the grounds. It had twenty-eight seats over two levels—making it highly social and very busy—and emptied into brick-lined drains which carried the waste to the River Thames. While latrines were popular, people still had the legal right and the social custom to excrete in the street or anywhere else they pleased, in plain sight of their community, indoors and out.

During the sixteenth and seventeenth centuries, new bodily habits, attitudes, and practices began to occur with the emergence of new instruction manuals, schoolbooks, and court regulations. In this regard, there are three issues highlighted by Erasmus's 1530 *De civilitate morum puerilium* worth discussing.

Example A

> It is impolite to greet someone who is urinating or defecating.[17]

In this example we glean at least two important distinctions: First, it would not have been an extraordinary occurrence to meet someone in your path or in your home who was urinating or defecating. Second, the need to issue this advice suggests that at some point in the not too distant past it was not considered impolite to speak to someone who was openly engaged in excretion, but rather that it was a relatively new social problem which required direct behavioural instruction.

Example B

> A well-bred person should always avoid exposing without necessity the parts to which nature has attached modesty. If necessity compels this, it should be done with decency and reserve, even if no witness is present. For angels are always present. . . . If it arouses shame to show them to the eyes of others, still less should they be exposed to their touch.[18]

Here we see an early attempt to instil bodily shame (through 'modesty') in the readers of this advice. Those receiving this instruction learned that they were *supposed* to start hiding their bodies and *begin feeling* ashamed not only while around other people but also while alone. Clearly, this modesty didn't come from within individuals, but instead was instituted upon a community in the name of 'nature' and 'decency', and a new level of morality in regard to the religious notion of ever-present angels. The instructions in examples A and B prop each other up and work to cooperatively *construct* bodily shame and modesty from many angles. They encourage people to consider questions pertaining to how and when they interact with others, how one's own body is viewed by others, and how to regard other people's bodies as potential sources of shame.

Example C

> To hold back urine is harmful to health. [And regarding farting or
> passing gas:] If it can be purged without a noise that is best. But it is
> better that it be emitted with a noise than that it be held back.[19]

Early concerns over health and illness as related to how one uses one's
body are introduced in this example. The conflict here is revealing. On
the one hand, there is a general understanding that it is not healthy to
hold in one's bodily functions, as it is widely believed to easily lead to
illness.[20] On the other hand, there is tension and delicacy about *how*
one should engage their bodily functions while in the presence of others. Still, at this time it was clearly more important to allow the bodily
functions to occur in public, in order to maintain the *appearance* of
health,[21] than to institute complete control over them. As explained
alongside example C: 'It is not pleasing, while striving to appear urbane,
to contract an illness.'[22]

It is vital, when considering these examples from a contemporary
Western perspective, to recognise the mere fact that these topics were
discussed at all and the openness with which they were discussed. As
Elias explains:

> The thoroughness, the extraordinary seriousness, and the complete
> freedom with which questions are publicly discussed here that have
> subsequently become privatised to a high degree and overlain in
> social life with strong prohibitions shows particularly clearly the shift
> of the frontier of embarrassment and its advance in a specific direc-
> tion. That feelings of shame are frequently mentioned explicitly in
> the discussion underlines the difference in the shame standard.[23]

As shame and modesty increased and began to be embodied from a
young age, it became increasingly inappropriate to speak directly about
these matters. For example, it would be shocking and 'inappropriate' to
see instructions in a contemporary schoolbook explaining the 'right' and
'wrong' way to use and feel about one's body during excretion; instead
this instruction now happens in early childhood as part of the toilet
training and socialisation process in general. It is something that is dealt
with in the *privacy* of the family home (not in the *public* domain of the
school) and it is seen chiefly as the responsibility of parents and guar-
dians.[24]

Along these lines, the continuing advancement of the modern excretory shift is evident in examples from only a slightly later period, namely
the sixteenth and seventeenth centuries. The following three examples
expose new societal attitudes and practices in Western Europe.

Example D

> [I]t does not befit a modest, honourable man to prepare to relieve
> nature in the presence of other people, nor to do up his clothes
> afterward in their presence. Similarly, he will not wash his hands on
> returning to decent society from private places, as the reason for his
> washing will arouse disagreeable thoughts in people. [25]

This example shows how modesty was pinned to a whole series of bodily
actions undertaken before and after excretion, with particular emphasis
on not disturbing other people who may be present. This imperative
required people to become increasingly cognisant of their actions *as
they were happening* and then have the wherewithal and foresight to be
able to change or modify their behaviour without social awkwardness
while in the presence of others. This set of psychic and physical practices is, again, something that individuals in contemporary society learn
during toilet training and early socialisation, but for adult individuals in
the sixteenth and seventeenth centuries, it meant consciously learning a
new way of being embodied. The obvious correlate to this new 'etiquette' is that people were then *expected* to be easily and instantly
disgusted (another learned response) by the *mere thought* of someone
else's bodily functions. This expectation is in stark contrast to example
A, which urged people not to *greet* someone who was *engaged in excretion*. The stunning difference between these two instructions, these
new social expectations, exposes the advancing desire to distance bodies
from one another. The instruction in this example is concerned with
keeping individuals unaware of the bodily needs of others, and it only
worked by keeping individuals hyper-aware of their *own* bodily needs
and actions so that they could be 'properly' managed and controlled.

In example D we also begin to see another form of bodily distancing
through the employment of euphemism. In this case the phrase 'relieve
nature' is taking the place of specific names of bodily waste and bodily
functions that were explicitly discussed and written about in European
contexts just twenty years prior. One additional point to consider in
example D is the instruction not to wash one's hands. This is interesting, as less than two hundred years later we see the exact opposite
instruction based on the same reasoning. In contemporary Western
societies individuals must at least make it *appear* as though they've
washed their hands after they've urinated or defecated to maintain hygienic decency and, for many, hand washing has become habituated
into their public toilet practices.

Example E

> Let no one, whoever he may be, before, at, or after meals, early or late, foul the staircases, corridors, or closets with urine or other filth, but go to suitable, prescribed places for such relief.[26]

This example is concerned with monitoring other people's bodies and actions. It is an encouragement for those hosting guests to instil propriety about where their visitors should be allowed to excrete, while also exposing all of the newly 'incorrect' places that people would normally go to. It is worth noting that we also see the continued use of euphemism in this example. Instead of 'defecate' or 'excrement' we now have 'foul', 'filth', and 'relief'.

Example F

> The extreme heat is causing large quantities of meat and fish to rot in them [the streets of Paris], and this, coupled to the multitude of people who . . .[27] in the street, produces a smell so detestable that it cannot be endured.[28]

Example F is unique insofar as it is part of a private correspondence written by a French woman, the Duchess of Orleans. She is writing about her visit to Paris and what is most intriguing is the ellipsis that appears. It seems as though where the Duchess could not write any words to describe the acts of human excretion she was witnessing, she wrote an ellipsis instead. She does not seem concerned with the *sight* of such acts, as she only reports on the smells produced by them. With the emergence of euphemism seen in previous examples, the words necessary for the description of her experience at the time of this correspondence have become unutterable and unwritable (and unreadable).

In all of the aforementioned examples (A–F), spanning over 160 years, we can see the beginning of the modern excretory shift: the early efforts to impose shame and modesty (in the name of 'decency', 'morality') onto natural bodily functions through psychic, linguistic, and physical means. Over the next few centuries the blurring, distancing, and 'quieting' of the toileting body increased. The advancement of the modern excretory shift means continued transformation of individuals within the West. The second and third spatial-historical milestones have a distinct feature not yet explicitly dealt with in the nearly 200 years already discussed: Gender difference, or sex segregation. This was a newly important distinction that required the erection of new spaces and advanced technology.

THE SECOND SPATIAL-HISTORICAL MILESTONE

The second spatial-historical milestone that represented an important move towards the contemporary public toilet spaces we rely on was the appearance of sex-segregated toileting spaces in 1739, initiated by the French upper classes. A restaurant in Paris, which was hired to hold a great ball, allotted toileting '*cabinets* with inscriptions over the doors, *Garderobes pour les femmes* and *Garderobes pour les hommes*, with chambermaids in the former and valets in the latter.'[29] This appears to have been the first time that separate toileting spaces were provided for men and women. While more precise details about the physical layout of these temporary spaces remain unknown, they were clearly intended to implement a gendered propriety and, as Sheila Cavanagh argues, to 'accentuate sexual difference and to project [that] difference onto public space.'[30]

At the time of the appearance of these new sex-segregated spaces, the continued quieting, shaming, and covering of the body and natural bodily functions further propelled the modern excretory shift. There was an increased emphasis on the visual, with strict orders to cover one's entire body, not only the sexual/excretory organs, from all eyes including one's own. By this time concern about bodily sounds and 'proper' language was increasing, as evidenced by the invention of even more euphemisms. To further explicate this phase of advancement I consider two eighteenth-century examples as laid out in *The Civilizing Process*. Examples G and H are from the same chapter ('On Parts of the Body That Should Be Hidden, and on Natural Necessities') but from two different editions of *Les règles de la bienséance et de la civilité Chrétienne* (The Rules of Christian Decorum and Civility) by La Salle (which was used in Christian schools). Example G is from the 1729 edition and example H is from forty-five years later, the 1774 edition. I will discuss these two examples together.

Example G

> It is a part of decency and modesty to cover all parts of the body except the head and the hands. You should care, so far as you can, not to touch with your bare hand any part of the body that is not normally uncovered. . . . You should get used to suffering small discomforts without twisting, rubbing or scratching . . .
>
> It is far more contrary to decency and propriety to touch or to see in another person particularly of the other sex, that which Heaven forbids you to look at in yourself. When you need to pass water [urinate], you should always withdraw to some unfrequented place. And

it is proper (even for children) to perform other natural functions where you cannot be seen.

It is very impolite to emit wind [gas] from your body when in company, either from above [burp] or from below [fart], even if it is done without noise; and it is shameful and indecent to do it in a way that can be heard by others.

It is never proper to speak of the parts of the body that should be hidden, nor of certain bodily necessities to which Nature has subjected us, *nor even to mention them.*[31]

Example H

It is part of decency and modesty to cover all parts of the body except the head and hands.

As far as natural needs are concerned, it is proper (even for children) to satisfy them only where one cannot be seen.

It is never proper to speak of the parts of the body that should always be hidden, or of certain bodily necessities to which nature has subjected us, or even to mention them.[32]

We can see, just by looking at examples G and H, that the ability to openly discuss bodily matters shifted considerably in just forty-five years. Example H is much shorter in length than example G though they cover roughly the same material and comprise the same chapter in their respective editions. As Elias highlights, while these two examples constitute the same chapter in name and topic, the 1729 edition (example G) 'covers a good two and one-half pages' and the 1774 edition (example H) covers 'scarcely one and one-half.'[33] By comparing these two examples, we can safely assume 'much that could be and *had to be* expressed earlier is no longer spoken of.'[34] For example, there is just a brief mention of sexual difference in example G, which does not appear in example H. The similitude of adult and child standards is also significant in these two examples. This is one of the first times that the bodily actions of children are held to the same high standard as adults, and if we consider the ever-rising levels of shame and embarrassment associated with adult bodies this is not surprising. By the eighteenth century, parents were not only expected to feel shameful about their own bodies, but also be embarrassed by the bodies of their children. By instilling these negative emotions in their children, parents could hope to avoid being betrayed and disgusted by them in the future.

There is one more point I want to make regarding examples G and H. When we compare these two examples to examples associated with the first spatial-historical milestone we can see that there is a clear shift from the imperative to manage one's body in accordance with one's own needs (e.g., to avoid illness), to being managed in line with a social standard or a moral 'decency'. In example G, the passage that begins 'It is very impolite to emit wind from your body' conveys the exact opposite advice given two hundred years prior, as detailed in example C.[35] To be considered a 'decent' human being in the eighteenth century one must control one's body entirely, for the sake of others and even when one is alone. The threat of illness occurring from within one's body as a result of holding in bodily waste, emphasised less than two hundred years prior, has now been entirely replaced by a concern for social propriety and moral indecency. People are not only told to learn to suffer bodily 'discomforts', but they are also encouraged to begin thinking about and experiencing their bodies from the outside in, actively disconnecting from their bodily sensations, desires, and needs. This is arguably an explicit sign of the emergence of individuality. Taken together, examples G and H emphasise the continued public 'modesty' (i.e., private shamefulness) expected of individuals and the move away from managing and *valuing* one's own bodily needs to managing the body according to the social and moral standard.

THE THIRD SPATIAL-HISTORICAL MILESTONE

The third and final spatial-historical milestone is the permanent, gender-segregated public toilet from the Victorian era. The emergence of this new space resulted from the six-month-long Great Exhibition of 1851 that was held at the Crystal Palace (erected in London's Hyde Park). This exhibition boasted the first public toilet facilities (with basically the same flush technology we use today), which were used by 827,280 visitors[36] who paid to use them.[37] According to Greed, at the beginning of the exhibition toilets were only provided for men, but, realising their error, the Royal Society of Arts (RSA) quickly arranged to provide more that women were permitted to use.[38] Possibly the planners of the Great Exhibition were expecting families, or at least women, to attend but were unable to anticipate the need feminine bodies would have to urinate and defecate in a public space.

After the success of these public toilets at the Great Exhibition, the RSA embarked on a scheme to design and implement a series of free-standing public toilets in London.[39] The timing and order of events that led to the implementation of the public toilets is unclear. This lack of clarity suggests how indirectly the process was dealt with (which is unsurprising considering Victorian attitudes towards 'delicate' matters of the body, waste, and hygiene), a point I expand below. According to Greed it was a relatively easy and quick process that occurred in just a matter of months. She reports that the first permanent public toilets (seemingly for both men and women, or rather, men and all 'others') opened in late 1851 and 'were supplied with a superintendent and two attendants each, and comprised two classes of toilets, for gentlemen and the masses.'[40] Shortly thereafter, in February 1852, a separate and specific public toilet opened for women, which was soon followed by a separate and specific public toilet for men. According to this estimation, the gender segregation at work here at first wholly othered all bodies (and thus those bodies' needs) which were not male and then later created a clear binary of men and women. Maybe, upon later contemplation, the men making the decisions regarding these spaces felt women *needed* to be kept separate from potential dangers of bodies that were not granted full male status.

However, the somewhat more plausible timeline, as interpreted by Kira, found that the process was prolonged by a few years and was not without obstacle. While there was a greater development of public toilets in the newly constructed rail terminals in the years immediately following the Great Exhibition, the first freestanding public toilets (i.e., facilities conceived of and built solely for the purpose of human excretion) in London were initially met with trouble. George Jennings, the man responsible for the public facilities at the Great Exhibition, proposed building '"halting stations" at strategic locations around London'. His offer (which included 'supplying and fixing the appliances free of charge') was rejected by the authorities.[41] As Jennings explains:

> [The] Gentlemen (influenced by English delicacy of feeling) . . . preferred that the Daughters and Wives of Englishmen should encounter at every corner, sights so disgusting [i.e., human excrement] to every sense, and the general public suffer pain and often permanent injury rather than permit the construction of that shelter and privacy now common in every other city in the world.[42]

At this point bodily functions, while still a feature of the public landscape (i.e., with excrement literally on the streets), were not considered a public issue, and thus public facilities were not something the authorities were willing to discuss at length or spend money on. Seemingly, the main obstacle hindering the Victorians' progression and urban development at this moment in time was their own attitude 'influenced by English delicacy of feeling'[43] about the role of the 'private' body in public life. Perhaps the authorities felt that if everyone adhered to the level of social propriety they did, public toileting would not be an issue, as everyone would deal with their bodily functions in private.

At this point in the modern excretory shift the discourses surrounding the body and bodily functions, as elucidated in several previous examples, were severely limited and shrouded in euphemism. As the Victorians were known for taking propriety to the extreme, it was easier and more socially acceptable among the elite to allow the somewhat medieval status quo to continue rather than have serious and direct conversations about the need for permanent excretory facilities. 'The subject was indelicate, and the problem was not admitted to exist.'[44] This exposes a conflict between the Victorian desire for an ever-increasing individual, self-managed, and quiet bodily etiquette and the willingness to then break this ideal etiquette in order to implement an infrastructure through which this bodily management could occur. Put simply, it was important to tell people how to manage their bodies and then expect them to individually deal with it themselves, but when it came to taking responsibility for the new bodily standard writ large it was not immediately seen as a societal problem. For the time being it was better left unsaid. Daily encounters of human excrement and common physical suffering and discomfort (from holding in waste) was bearable and honourable but publicly admitting that they, the decision-making Gentlemen of the time (and no doubt their daughters and wives too), partook in the private and disgusting acts of excretion witnessed all around them was utterly unbearable.

Jennings won the toileting battle and 'went public' in the 1870s when public toilets were deemed necessary and shortly thereafter became widely available.[45] According to Greed, 'The first permanent public toilets for women, and men, were built in 1893 opposite the Royal Courts of Justice in the Strand, London.'[46] It seems the change came only when the Gentlemen of the time could no longer ignore the issues being caused by the lack of facilities and realised the conversation was worth having.[47] With Victorian delicacy suppressed long enough to make the vital decision regarding the need for public toilets, we arrive at the final spatial-historical milestone: the first gender-segregated, permanent, flushing public toilet facilities. These modern toilets, and the embodied practices of their time, are the not-so-distant relatives of the public toilets operating in much of the Western world today. This important development resulted in a new public social space that surely required a specific set of manners and etiquette, but there is little documented instruction on the proper way to use them.

Considering the unspeakable nature of bodily functions, it is not surprising that we find little discussion or social instruction regarding these new public spaces by the nineteenth century. Despite the new public space, the bodily matters those spaces were built for were no longer in the realm of the public, but rather something dealt with in private. Bodily restraint, silence, and individuation choked the topic, and the Victorians used social taboo and the guise of health to continue the trend. While health concerns are not primary in most of the examples above 'in the 19th century . . . they serve as instruments to compel restraint and renunciation of the gratification of drives.'[48] As shown throughout the preceding examples, individuals learned self-restraint through mostly social and moral pressures, whereas Victorian society pushed this further by directly employing moral threats via 'health' and 'hygiene' as the way to demand self-control, suppress pleasure, and further the specific emotional modelling of individuals. As Elias explains:

> These hygienic reasons . . . played an important role in adult thinking about civilisation, usually without their relation to the arsenal of childhood conditioning being realised. It is only from such a realisation, however, that what is rational in them can be distinguished from what is only seemingly rational, i.e., founded primarily on the disgust and shame feelings of adults.[49]

The overall process of the modern excretory shift, including increasing levels of shame, embarrassment, modesty, disgust, and repugnance cannot be fully understood by exploring in isolation the development of bodily practices or the advancement of technologies (e.g., flush toilets, sewage systems). The social changes that occurred from the sixteenth to the nineteenth century have since persisted and can only be understood through the combination of these factors. It would be very difficult to understand the societal shift only by looking at the scientific discoveries and technological inventions. Instead, when we consider the societal interventions and the overall transformation of bodily needs, it becomes evident that 'the development of a technical apparatus corresponding to the changed [social] standard consolidated the changed habits to an extraordinary degree. This apparatus served both the constant reproduction of the standard and its dissemination.'[50]

In the sixteenth and seventeenth centuries open bodily excretion was taken for granted as part of everyday life much in the same way that we, the contemporary products of the modern excretory shift, take for granted bodily control, self-restraint, and the availability of flush toilets in our daily lives (and surely the absence of cesspools and human excrement in our streets). This process can be seen as playing an important part in transforming the open body into a closed or bounded body (an understanding and related style of embodiment that persists today as *homo clausus*). In order to fully understand the history of modern public toilets I will lastly consider the dimensions of the modern excretory shift that have most contributed to the production of individuality, the closed-rational body, and the relationship of publicness and privateness.

TO ARRIVE AT PUBLICNESS IS
TO BEGIN AT INDIVIDUALITY

With the techno-spatial-historical timeline clear, it is vital to return to a point introduced earlier, the concept of 'publicness'. Publicness and privateness are central to the development of public toilets, as they are ideological systems that helped shape individual desire, modesty, and guilt in relation to how one should feel 'in' and use their body in everyday life. The transformation from the beginning of this account in the fifteenth century to the late nineteenth century, through the three specific milestones and the broader, more subtle and entangled bodily and interpersonal changes (taken together, I term the modern excretory shift), can be generally understood as a transformation of how people related to their own body and to one another generally. This can be characterised as contributing to a shift from a *gemeinschaft*, or community of people, to a newly formed *gesellschaft*, or society. As a result of this civilising process there was a distinct transformation from a communally based culture to a society of individuals, and it is my suggestion that the formation of the modern toileting body played a large, albeit largely unacknowledged, part in that shift. The *gesellschaft* resulting from the modern excretory shift is made up of individuals who harboured shame and embarrassment about their bodies and bodily functions that were once comfortably expressed in public, often as a social activity, without fear or taboo. The process of individuation included shaping people's thoughts and feelings about their bodies, creating new practices of embodiment, and developing new spaces and technologies specifically for the acts of excretion. Through these related changes, people took on what David Inglis terms the 'bourgeois faecal habitus'. He explains that the

> repressions over excretory practices were dependent upon alterations in the ways that the human body in general terms, and that's also in excretory terms, was received and represented. Repressions over practices were thus dependent upon a shift from the set of representations of the body's excretory capacities that were dominant in the medieval period, to those informed [by the] bourgeois faecal habitus.[51]

A catalyst for the change in the way that people viewed and used their bodies, from the medieval to the Victorian period, resulted 'from the development of class habituses over time.'[52] The bourgeois faecal habitus grew out of general attitudes regarding bodily restraint and repression. As Bourdieu has identified more specifically, it was reliant upon class distinction: 'The denial of lower, coarse, vulgar, venal, servile—in a word, natural—enjoyment, which constitutes the sacred sphere of culture, implies an affirmation of the superiority of those who can be satisfied with the sublimated, refined, disinterested, gratuitous, distinguished pleasures forever closed to the profane.'[53] Therefore, by the late nineteenth and early twentieth centuries, as a result of the emerging hierarchy in class structures,

> the ruling classes were obsessed with excretion. Faecal matter was an irrefutable product of physiology that the bourgeois strove to deny. Its implacable recurrence haunted imagination; it gainsaid attempts at a decorporealization; it provided a link with organic life. . . . The bourgeois projected onto the poor what he was trying to repress in himself. His image of the masses was constructed in terms of filth. The fetid animal, crouched in dung in its den, formed the stereotype.[54]

As this attitude was projected onto the lower classes by the ruling classes, the lower social classes were compelled to adopt the habitus of those above them, and as it was instituted as a *moral* imperative, they changed their bodily practices accordingly. 'For the bourgeois faecal habitus, bringing moral cleanliness to the city, and to the proletarians who dwelled therein, involved an 'orderly' recasting of urban space.'[55] Vital to this new urban space was the distinction of publicness and privateness.

The concept of 'publicness' draws on a confluence of factors. As it directly relates to public toilet spaces, those factors include: 'the degree of strangeness of other users from oneself, the extent of usage of a facility, and perhaps most important ultimately, the level of cleanliness and maintenance, which, in turn, relates to our concerns regarding territoriality and privacy.'[56] These factors did not exist at the beginning of the modern excretory shift (in the early fifteenth century) and neither did such a consolidated concept of 'privateness' or public toilets. It is only through the development of the modern excreting body and the ultimate dominance of the bourgeois faecal habitus, that these concerns were developed, made relevant, and began to seep into everyday thought. Bodily concerns about privacy did not pre-exist the development and adoption of public toileting practices. Directly related to and reliant upon 'publicness' is the sense of 'privateness'. Privacy (and the feeling of privateness it enables) is significant here because, as previously demonstrated, it is a learned social value that has become necessary in order to maintain socially acceptable behaviour. As Kira explains, 'we must have privacy in certain instances so that we do not violate cultural norms specifying that certain things be done in private.'[57] What's more, 'privacy and privateness sustain . . . our sense of individual identity' and 'in its simplest form it involves "aloneness", or freedom from the presence and demands of others.'[58] Similarly, in Goffman's terminology, privateness is directly related to our 'territories of the self' and the ability to maintain our public performances.[59] Considering these points together, we arrive at the paradoxical nature of public toilets.

While public toilets are spaces where we can seek privacy from the masses for our 'disgusting', unspeakable acts, they are not private enough to shield us from the 'disgusting', unspeakable acts of others. So as long as there is the continued threat that someone else may see or hear us while engaged in excretion, the Victorian taboo, we are bound to the rules and codes of the space and the bourgeois faecal habitus that honours 'cleanliness' and repression. What's more, 'while our own excretory processes and products may be more or less disagreeable,' as long as the processes are carried out in the socially prescribed ways and we never actually have to come into direct contact with their products, 'those of strangers tend, in general, to be viewed even more negatively.'[60] Just as the spaces themselves maintain the Victorian ideal, this social standard of considering other people's bodies and bodily waste as alien or threatening continues the sense of self as individually bound and entirely separate from other bodies. This way of being denies the entangled nature of bodies, *through the experience* of their entanglement. That is to say, it is only through the sensorial impact other excreting bodies have on our own that we can feel a particular way about them as alien or other. Therefore individuality, as a way to maintain the borders of oneself, became increasingly important throughout this civilising process.

NOTES

1. Lucinda Lambton, *Temples of Convenience and Chambers of Delight* (Stroud, UK: Tempus Publishing, 2007), 95.

2. Rose George, *The Big Necessity: The Unmentionable World of Human Waste and Why It Matters* (New York: Metropolitan Books, 2008), 52–53.

3. Sheila Cavanagh, *Queering Bathrooms: Gender, Sexuality, and the Hygienic Imagination* (Toronto: University of Toronto Press, 2010), 84.

4. This text was problematic for booksellers because of the 'uncouth' content and the photos of nude bodies (though with black-barred eyes) and was therefore categorised as 'dirty' and 'pornographic' and/or 'comedy' until it went out of print. It is now a very difficult text to locate.

5. Alexander Kira, *The Bathroom* (New York: Viking Press, 1976), 5.

6. It has also been reported that 'primitive forms of the flushing toilet, together with channels to carry foul water away, were found at the 3,700-year-old palace of King Minos at Knosses.' George, *Big Necessity*, 24.

7. Cavanagh, *Queering Bathrooms*.

8. Clara Greed, *Inclusive Urban Design: Public Toilets* (Oxford: Elsevier, Architectural Press, 2003), 32.

9. Cavanagh, *Queering Bathrooms*, 28. Also see Greed (1995) and Penner (2005).

10. Kira, *Bathroom*.

11. Kira, *Bathroom*, 6.

12. Lawrence Wright, *Clean and Decent: The Fascinating History of the Bathroom and the Water Closet, and of Sundry Habits, Fashions and Accessories of the Toilet, Principally in Great Britain, France, and America* (London: Routledge, 1963), 102.

13. Greed, *Urban Design*, 34.

14. Norbert Elias, *The Civilizing Process: Sociogenetic and Psychogenetic Investigations* (Oxford: Blackwell, 2000), 119.

15. This process and its corresponding socio-bodily attitudes and practices is later referred to as the 'modern excretory shift' and is concerned explicitly with the changes that occurred starting in the sixteenth century and continued through the nineteenth century.

16. Kira, *Bathroom*, 167.

17. Erasmus 1530, in Elias, *Civilizing Process*, 110.

18. Erasmus 1530, in Elias, *Civilizing Process*, 110.

19. Erasmus 1530, in Elias, *Civilizing Process*, 110.

20. There was clearly a popular stance, as Erasmus explains: 'Regarding the unhealthiness of retaining the wind: There are some verses in volume two of Nicharchos' epigrams where he describes the illness-bearing power of the retained fart, but since these lines are quoted by everybody I will not comment on them here.' Erasmus (1530) in Elias, *Civilizing Process*, 110, 111.

21. While it was still socially acceptable to pass gas in public during this time, it was ultimately allowed in order to maintain the appearance of good health, which was still a social end. As the body becomes more intensely divided and heavily controlled, audibly passing gas later becomes a sign of illness—or at least illness is often invoked when it happens in public.

22. Erasmus 1530 in Elias, *Civilizing Process*, 110.

23. Elias, *Civilizing Process*, 111.

24. Some of the participants in this study who were privately employed by busy New York City families partook in the toilet training of the children they

were paid to care for as they typically spent more time with the children, on daily and weekly bases, than parents spent with them.

25. From *Galateo: or, A Treatise on Politeness and Delicacy of Manners*, Della Casa, 1558, p. 32 in Elias, *Civilizing Process*, 111.

26. Brunswick Court Regulations 1589 in Elias, *Civilizing Process*, 111.

27. This ellipsis appears in the original text to serve as a 'blank' or to indicate the absence of an appropriate word.

28. Duchess of Orleans 1694 in Elias, *Civilizing Process*, 112.

29. Wright, *Clean and Decent*, 103, original emphasis.

30. Cavanagh, *Queering Bathrooms*, 28.

31. La Salle 1729, p. 45ff. in Elias, *Civilizing Process*, 112, my emphasis.

32. La Salle 1774, p. 24 in Elias, *Civilizing Process*, 113.

33. Elias, *Civilizing Process*, 113.

34. Elias, *Civilizing Process*, 113, my emphasis.

35. Elias, *Civilizing Process*, 113.

36. Regarding the wild popularity of the public (pay) toilets at the temporary Crystal Palace, the Official Report after the Great Exhibition stated that: 'No apology is needed for publishing these facts, which . . . strongly impressed all concerned . . . with the sufferings which must be endured by all, but more especially by females, on account of the want of them.' Despite this, 'when the building was re-erected . . . it was strongly urged on grounds of economy that lavatories should be excluded.' This battle was ultimately won by Jennings, who 'was crusading for the provision of "conveniences suited to this advanced stage of civilisation" in place of "those Plague Spots that are offensive to the eye, and a reproach to the Metropolis".' Wright, *Clean and Decent*, 200.

37. Kira, *Bathroom*, 195; Wright, *Clean and Decent*, 200.

38. AMC, Newsletter, January (1997) in Greed, *Urban Design*, 42.

39. Greed, *Urban Design*, 42.

40. Greed, *Urban Design*, 42.

41. Kira, *Bathroom*, 196; Wright, *Clean and Decent*, 200.

42. Jennings in Wright, *Clean and Decent*, 201.

43. Jennings in Wright, *Clean and Decent*, 201.

44. Wright, *Clean and Decent*, 200.

45. Wright, *Clean and Decent*, 201; Kira, *Bathroom*, 195.

46. Clara Greed, 'Public Toilet Provision for Women in Britain: An Investigation of Discrimination against Urination', *Women's Studies International Forum*, 18, no. 5 (1995), KL 824–25. It is difficult to determine when the first public toilets were actually built and opened. This difficulty is most evident, perhaps, when the same scholar—in this case Clara Greed—reports two different dates for the same happening.

47. I assume that this must have been connected to cholera outbreaks and contaminated water systems, though I haven't found this link written about explicitly.

48. Elias, *Civilizing Process*, 115.

49. Elias, *Civilizing Process*, 114–15.

50. Elias, *Civilizing Process*, 119.

51. David Inglis, *A Sociological History of Excretory Experience: Defecatory Manners and Toiletry Technologies* (Lewiston, NY: Mellen, 2001), 116.

52. Inglis, *Excretory Experience*, 16.

53. Pierre Bourdieu, *Distinction: A Social Critique of the Judgement of Taste* (Cambridge, MA: Harvard University Press, 1984), 7.

54. Alain Corbin, *The Foul and the Fragrant: Odour and the French Social Imagination* (Cambridge, MA: Harvard University Press, 1986), 144.

55. Inglis, *Excretory Experience*, 234.

56. Kira, *Bathroom*, 201.

57. Kira, *Bathroom*, 168.

58. Kira, *Bathroom*, 166–67.

59. Erving Goffman, *Relations in Public: Microstudies of the Public Order* (New York: Basic Books, 1971).

60. Kira, *Bathroom*, 201.

III

Theory as Practice

5

HOMO CLAUSUS AND THE NORMATIVE INTRA-ACTION ORDER OF PUBLIC TOILETS

Abjection . . . is immoral, sinister scheming, and shady: a terror that dissembles, a hatred that smiles, a passion that uses the body for barter instead of inflaming it, a debtor who sells you up, a friend who stabs you.

—Julia Kristeva

This chapter seeks to locate *homo clausus* ways of being in public toilet spaces. As explored in previous chapters, *homo clausus* is an onto-epistemology that permeates much of the Western philosophical tradition's understanding and construction of individual selfhood. *Homo clausus* is both a mode of embodiment and a system of constructing/comprehending the world that is predicated upon an immense amount of ongoing social, emotional, physical, and mental work concerned with shoring up the boundaries of the individual self; the process is conceived to be directed by one's independent mind from within one's bounded body. The management and maintenance of *homo clausus* self-body relationships take various forms, via different sets and styles of material-discursive practices, which are (generally) socially instituted and culturally circumscribed. The social and personal processes employed within English-American public toilet spaces are no exception. As evidenced in my empirical data, the particularly *homo clausus* practices repetitively employed within public toilets can be understood as condensing around the three primary features of dis-embodiment outlined in chapter 1: Socialisation, rationalisation, and individualisation. In order to focus on the material-discursive practices that are based upon these three bodily characteristics of *homo clausus* identity, I have broadly categorised the communicated behaviours, ideas, and experiences of those interviewed for this study into three rules or phenomena that amount to what I term the triadic intra-action order (TIO). The rules of the TIO are as follows:

1. Minimise your movement (socialisation)
2. Mind your eyes (rationalisation)
3. Manage your boundaries (individualisation)

I use the term *TIO* in order to refer to a set of practices that are embodied from a young age and that, through repetition, appear natural and universal, not taught and learned. This order takes natural bodily functions and in certain circumstances, that is, in the presence of others in a public toilet space, imposes upon them a rigid, highly controlled system of bodily management in order to maintain identity. The TIO takes something natural and universal, and renders it dangerous, threatening, abject. As Butler observes, these constructed identities and related bodily dispositions appear to be based upon an essence which give the identity substance, when it is only merely constituted: 'the appearance of substance is precisely that, a constructed identity, a performative accomplishment which the mundane social audience, including the actors themselves, come to believe and to perform in the mode of belief.'[1] Like Butler's brand of performativity, the TIO prohibits sensory-embodiment from playing an active role in experiential becoming, which has implications for understanding how *homo clausus* bodies are constructed, disciplined, and made fragile in daily life. Through the TIO, the body is rationally directed according to the learned principles of this mode of interaction, making it an excellent example of an onto-epistemology. What is most readily apparent in these spaces is how social power works 'from below' in the Foucauldian sense, 'regulating the most intimate and minute elements of the construction of space, time, desire, embodiment'.[2] As feminist philosopher Susan Bordo, interpreting Bourdieu, asserts, 'Banally, through table manners and toilet habits, through seemingly trivial routines, rules, and practices, culture is "made body" . . . converted into automatic, habitual activity.'[3] By understanding these daily habits of social life we can better understand how *homo clausus* dis-embodiment is lived out and reproduced through a set of sexed-gendered practices and performances that affect embodiment, desire, and public life. Before unpacking the rules, I will situate their general operational purpose in relation to the abject body.

Material-discursive practices that coalesce into these three categories differ for the respective sexed-gendered spaces (i.e., men's rooms and women's rooms), but generally serve the same purpose: to simultaneously perpetuate and assuage entangled feelings of bodily fear, anxiety, shame, and embarrassment (FASE), in an attempt to stabilise individual rational, monadic identity. These feelings within these spaces are tied intimately into what Kristeva terms the abject. The abject lurks in the spaces between one's body and one's mental body image, requiring one to rationally 'keep an eye' on what escapes the outer layer of skin— i.e., the imagined outer boundary of *homo clausus* individuality. It is, as Morris and Sandilands explain, 'the revolting outside to the body born from within it, without which its appearance of separateness is not possible'.[4] The abject are those bodily excretions that are part of one's body but with which one does not subjectively identify. What we have come to consider abject is directly related to *homo clausus* identity and the desire to maintain it. As Kristeva explains, 'It is thus not lack of cleanliness or health that causes abjection but what disturbs identity, system, order. What does not respect borders, positions, rules. The in-between, the ambiguous, the composite.'[5] The body's abjections and their associated connections to FASE help prop up the borders of and are considered a threat to the stabilisation of the monadic individual. Crucially, 'this threat is intensely, and nauseatingly, corporeal; mucus, blood, faeces, vomit, pus are expelled from the body as if they were alien. But they aren't, and they do not go away.'[6] What is abjected by *homo clausus* simultaneously exposes its failure to be contained and gives relief to the practices one employs in attempting to maintain the sense of a monadic self. Therefore, the TIO of English-American public toilets is a technology of *homo clausus* abjection which requires individuals, in order to remain socially viable, to engage constantly in actions and behaviours which seek to distance one's self from the abject. Bodily FASE act as constant reminders that the 'disgusting' corporeal body is at odds with one's rational mind; it maintains the stiff barrier between what is known rationally and what is experienced in sensory-embodied becoming. As Morris and Sandilands put it: 'Embodiment thus concerns a constant tension between the *impossible* image of the whole, clean, coherent body and its multiple abjections. At once physiological and social, abjection is part of the process of bodily materialization in which power ritually writes the body, organizing corporeal desire and disgust.'[7] In the subsequent elucidation of the intra-action order of public toilets I aim to show how bodies in their potentially open-ended and boundless capacities for becoming (*corpora infinita*) are actively converted via rationally driven sensory-embodied practices into conceptually closed and bounded forms. These practices are material-discursive because they implicate both the living of the body and the ways we rationally understand how bodies *should be lived.* In a broad yet basic sense, this is both the construction and mediation of desire and disgust insofar as material-discursive practices structure how the body can live in space with other bodies and how, when it doesn't, it is understood as disgusting. By starting with the premise that the body is always already actively open, and the *homo clausus* must work to rationally separate and close it, we can begin to identify how and when we can most usefully allow the material experience of sensory-embodiment to feed back into the discursive loop. That is a move which represents the potential to shift from dis-embodied being to sensory-embodied be-coming.

INTRA-ACTION

The TIO of public toilets is developed from Erving Goffman's 'interaction order'; making the observation that daily life for most people entails the direct presence of others.[8] Even when not in situations of visible contact with others, nearly everything we do is socially situated. Furthermore, Goffman recognises that the arrangement of gender-segregated public toilets 'is totally a cultural matter. And what one has is a case of institutional reflexivity: toilet segregation is presented as a natural consequence of the difference between the sex-classes, when in fact it is rather a means of honoring, if not producing, this difference.'[9] He also realises that within these spaces women must spend time in the company of other women, yet gives little attention to the time that men must also spend in the company of other men or to the intricacies of movement and negotiation of bodies that occur in these spaces.[10] In short, Goffman fails to explore how gendered interaction within these segregated regions may do the work of (re)producing gender difference and how the body is experienced according to that difference.

Developing Goffman's analysis of the interaction order, this chapter explores men's and women's public toilets by suggesting that there exists in these places an intra-action order which normalises and condenses users' actions into conventionally masculine or feminine identities. Goffman employed the term *interaction order* to refer to how individuals' ways of negotiating physical co-presence can be viewed as a system which seeks to minimise disruptions in social life and manage and maintain the social fabric.[11] His approach towards the interaction order is to treat its effects as 'indicators, expressions or symptoms of social structures', yet he intentionally displays 'no great concern to treat these effects as data in their own terms.'[12] By moving beyond Goffman in this respect, by treating these 'effects' *as empirical data*, I will show how the TIO of public toilets is not merely reliant upon or bound by

rules derived from biological necessity, but can be viewed as an important instrument for managing the threat that leaky, naked, unstable
bodies pose to conventional constructions of *homo clausus* identity. In
contrast to Goffman's suggestion that there exists a universal 'interaction order', based upon certain common features of human embodiment, this chapter explores the ambiguous public-private space of gender-segregated public toilets and suggests they are characterised by a
distinctive 'order' of intra-action predicated upon a code of *homo clausus* dis-embodiment. I posit that the intra-action order is, rather than
universal and reflective of ordinary embodiment, a socially taught and
learned system of *homo clausus* monadic, rational individuality.

It is evident, for example, in Khadijah Farmer's story, one of those
collected for my research, that such norms exist in women's public
toilets and that transgressing these norms, even just through appearance, can be highly threatening. Farmer, a gay woman in her late twen-
ties, was thrown out of a West Village restaurant in New York City, after
the Gay Pride Parade in 2007. As reported on the *New York Times* City
Room blog, Farmer explained the situation:

> He [the security guard] began pounding on the stall door saying
> someone had complained that there was a man inside the women's
> bathroom, that I had to leave the bathroom and the restaurant. . . .
> Inside the stall door, I could see him. That horrified me, and it made
> me feel extremely uncomfortable. I said to him, 'I'm a female, and
> I'm supposed to be in here.' After I came out of the bathroom stall, I
> attempted to show him my ID to show him that I was in the right
> place, and he just refused to look at my identification. His exact
> words were, 'Your ID is neither here nor there,' which means that
> my ID didn't matter to him.[13]

This example is revealing. Here, what one person thought they saw
(seeing = knowing)—i.e., a man in the women's room—was evidence
enough for security personnel to take action and deny someone their
rights. Instead of listening to Farmer's voice, pausing to look at her, or
at the only thing she could use to defend herself, her state-issued ID,
the security guard had already decided on the set of actions he was
taking; rational, masculine, goal-oriented progression. One person's selective attention in determining Farmer was a man, by seeing only her
masculine characteristics, meant that she was then treated as a criminal,
as abject. In that moment of rational, selective attention her humanness
was taken from her and she was thrown out like a piece of garbage.
Clearly, the comfort of those who embody conventional *homo clausus*
gender expression and identity took precedence in this situation. The
fact that this occurred in the West Village, an extremely important

neighbourhood for the gay rights movement historically, and on the day of the Gay Pride Parade, show the problems that can confront individuals who transgress conventional gender-expression boundaries. Despite her identity as a woman, Farmer's unconventional femininity was seen as highly threatening to other, more conventionally feminine individuals. This threat is similar to the fear produced in transmen who use men's rooms. I spoke with many who, despite always passing (as 'real' men assumed to have a penis), still often fear being 'found out' (as not having a penis and thus not actually being 'real' men) and threatened with violence. This fear persists even though they have no personal precedent for such violence and could not use a women's room without trouble.

By examining the intra-actional practices within men's and women's public toilets, through the TIO I aim to begin bridging this gap in Goffman's writing; by demonstrating that what occurs ordinarily within these spaces involves a heteronormatively directed embodied reflexivity (i.e., men and women managing their bodies and their interaction with others on the basis of prevalent standards of *homo clausus* 'decency' underpinned by FASE, selective attention, and sensorial individuation) which is reproductive of hegemonic heterosexual masculinity and femininity. While they often provide people with a 'structure' on which to anchor identity, I view these attributes as limiting because, for example, they deny rigidly gendered bodies an openness or fluidity of exchange with other like-gendered bodies. As Sara Ahmed states: 'Heterosexual genders form themselves through the renunciation of the *possibilities* of homosexuality, as a foreclosure *which produces a field of heterosexual objects* at the same time as it produces a domain of those whom it would be impossible to love.'[14]

To remain stable and valid, the norms that imbue dominant notions of *homo clausus* heterosexuality rely on binary gender expression (i.e., men are men because they are not women), the embodiment of 'correct gender roles' (e.g., men are stronger, more powerful, and less emotional than women), and a corresponding (assumed) material reality of external (and thus internal) genital organs (i.e., masculine bodies have penises, testicles and feminine bodies have vulvas, breasts). Thus, I use the term 'hetero' throughout this chapter to do more than act simply as the first part of a term denoting a specific identity predicated upon belonging to one of two opposing and opposed sexes, it is more than a term which carries representational ways of being, it is an onto-epistemology. Thus, I am referring to a range of practices and ways of being that prop up specific assumptions of conventional, *homo clausus* gender and corresponding heterosexuality—i.e., how bodies are expected to be lived *based on* their materiality. For example, hetero men are rational and thus not physically or emotionally expressive in their mannerisms and style of communication, and because women's bodies are soft, delicate, and always smell good, hetero women do not enjoy engaging in activity or labour that results in sweating and getting 'dirty'. According to Saunders, 'inherent in such renderings of the body is an anti-body sentiment, which seeks to curb and control the body's openness to possibilities.'[15] Hetero is a mediation of the *homo clausus* body and the abject. It works to straighten, organise, and order bodies and with the TIO, as a technology of *homo clausus* abjection, it is possible to understand how this works not merely discursively, but through fleshy material, feeling, sensing bodies. What interests me in this chapter are the forms heteronormative embodiment takes in everyday life (at even the most subtle and intimate levels) and how in identifying those forms, we can better understand the reproduction of gender, desire, and sexuality as firmly bound within the rigid subject position of *homo clausus* identity.

METHODS

Using data from forty-five semi-structured interviews and over two hundred 'toilet use' surveys, I elucidate the triadic dimensions of the intra-action order of public toilets. The data were collected in roughly equal amounts in Southern England (London, Canterbury) and New York City. It is important to note that Goffman's observations are generally seen as relevant for English-American society and have been used in that context and, empirically, my results seem to be the same for both areas.[16] This is unsurprising because both cultures tend to construct basic, individual (i.e., *homo clausus*) identity in similar ways. Therefore the exact social practices and related socially instituted feelings may differ slightly between societies, but the underlying identity structure they seek to maintain remains the same. Additionally, I want to emphasise that I am dealing with bodily functions that, although culturally shaped in all manner of ways, are anthropologically common parts of what it is to be a human across all cultures and societies. Accordingly, I have analysed the data together in an effort to forge a cohesive English-American observation.

While not necessarily representative of the population as a whole, this sample is useful theoretically, as it serves to highlight the rule-bound nature of toilet use. Interviews with differently identified users of public toilet spaces offer the greatest opportunity of having the complex and deeply performative nature of the rules of these spaces brought to the fore. Trans (e.g., -gender, -sexual, or just 'trans') and queer (genderqueer or just 'queer') identities are particularly important in this study not only because of the status of their already non-conventionally bound identities, but also because these individuals are often most conscious of their bodies in these spaces (often not by choice). Many trans and queer individuals find that they are already breaching a rule of the action order by merely being present in the space (based on their appearance entangled with their sexuality—i.e., sex-gender-sexuality) when, generally, they just want to use the toilet in peace like everyone else. That is to say, just because certain individuals choose to identify and present outside of conventional (heterosexual) sex-gender-sexuality, it does not mean they are actively seeking to break any of the rules of the action order in their daily use of these spaces.

In what follows it becomes clear that each user of the space is expected to follow and reproduce the triadic dimensions of the intra-action order, thus differently identified users provide varying opinions, experiences, and insights. For example, the use of the mirror may be seen as natural and necessary for a heterosexual woman, whereas for a queer user, the mirror may be seen as a locale of uncomfortable pressure. It is through these differing experiences that I hope to show the inflexible nature of the TIO that works to maintain the most rigid and unnatural aspects of *homo clausus* dis-embodiment. Furthermore, I present the experiences of the sample together, rather than based upon the two spaces, to show how the TIO works on and through *homo clausus* bodies to create order and identity, via material-discursive practices, where it does not naturally coalesce.

RULE ONE: MINIMISE YOUR TIME AND SPACE (SOCIALISATION)

Once the *homo clausus* subject has determined where to locate the imagined borders of their body, through sensorial individuation and other early socializing processes, they must engage in practices which continually help maintain that sense of boundedness. Without the imagined borders of the body-self, there could be no subject and no object. Therefore, this is a process which helps manage who and what one is. One of the most fundamental ways of managing this dis-embodied state is simply how one's body can be in, use, move through, and take up space. Public toilet spaces are generally understood as goal-oriented spaces that people try to spend as little time in as possible, as they are imbued with an underlying anxiety and general fear of transgression (which may erupt into shame and embarrassment). Thus, users of men's and women's rooms alike are taught, when it comes to carrying out everyday bodily functions at an away-from-home toilet, to be as quick and direct as possible. While women may freely spend time in front of a mirror, a highly normalising activity I will explain in rule two, the time spent in a cubicle is strictly monitored. Similarly, while men may feel free to spend more time in a cubicle than women do, they are expected not to linger at urinals, sinks, or in front of mirrors. Therefore the management of the body in time and space may differ slightly for the gendered spaces, as those slight differences do the work of perpetuating sex-gendered difference, but on the whole this rule operates in nearly the same way for both spaces in an effort to maintain the sense of individual boundedness and the disconnection from other bodies. This sameness is characteristic of heteronormative, patriarchal gender, which is constructed along a singular axis and does not recognise sexual difference (i.e., one's specific materiality) as *ontological* difference. If the boundaries of the body are allowed to loosen, connections with other bodies may become much more fluid, entangled; a more proces-sual state which is considered antithetical to rational, *homo clausus* identity.

This attitude is reflected in the practices of economic movement. The economy of movement means that bodies move through space in a rational and directed fashion, wasting as little time, space, and motion as possible. When speaking about overarching regimes of patriarchal power, and evoking Foucault, feminist phenomenologist Sandra Bartky recognises, 'The body's time . . . is as rigidly controlled as its space.'[17] Here, the mind directs the body in a purposeful way; the body is a tool, a means to a rational end—one first looks and then moves directly to *that* urinal or *that* cubicle—which is, ironically, ultimately serving a physical, not a rational, end: excretion. Here rationality directly impli-cates embodiment and embodiment directly implicates rational pro-cesses. Natural bodily functions, especially in public, are inherently abject and thus threatening to the order and borders of mature, rational *homo clausus* identity. So, for example, rather than movement in and through a toilet space being understood or experienced as sensory-embodied, it is instead made into rational, directed progression. Mate-rial-discursive practices that support this rationale and help maintain the sense of a closed body—even while the body is engaged in acts of literal openness—are evidence in both my interview and survey data. When it comes to the 'different' gendered spaces, the rule functions in equal ways yet with opposite emphases. While users of both men's rooms and women's rooms are expected to monitor their bodies accord-ing to both time and space, there is greater emphasis for users of women's spaces to be concerned about time and for users of men's spaces to be concerned about space.

As women's cubicles are often oversubscribed, users who confront this situation have no choice but to wait in line; a situation that often causes them to be hyperaware of the time spent within a cubicle and of how they use their bodies in space. As Plaskow points out, 'Women's willingness to wait on line offers important insights into the process of female socialization' and while not as overt as a school bell for example, the underlying pressure to manage one's body in terms of how much time is taken is a system of disciplining the body material-discursively.[18] This waiting increases the numbers of closely present bodies, and subjects those users actually occupying cubicles to the potential monitoring of near present others, a point I will come back to in rules two and three. What exactly are they doing in there, what are they 'admitting' to doing based on the amount of time being taken, and are they legitimately denying access and use to other users in making other women wait in line? This monitoring of absent yet also present bodies in enclosed toilet spaces can be seen as highly discomforting, destabilising, and threatening to conventional femininity, as it is often associated with bodily FASE. In order to maintain *homo clausus* boundaries and manage bodily FASE (by keeping it present yet under control), users of women's public toilets are concerned with minimising their time spent in a cubicle. For example:

> Lucy, twenty-three years old, bisexual woman: I'm more interested in getting in and getting out as quickly as possible. In and out!
> Billie, twenty-three, genderqueer: I try to go as fast as possible.
> Cece, twenty, heterosexual woman: I always feel rushed. . . . Just because I want to get out really quickly so I consciously don't spend a lot of time in there.
> Natalie, twenty-four, queer: I try not to spend a lot of time in there.
> Elizabeth, twenty-four, queer: I feel rushed and don't like to keep people waiting; this definitely causes anxiety in me.

Each of the above quotes was made in the context of longer comments referring to the users' discomfort at spending any longer than necessary in what they viewed as an identity-threatening locale. The pressure to spend as little time as possible limits how individuals can use their bodies. For example, Alice, twenty-six, who identifies as a queer female, but not as a 'woman', explains how this pressure requires her to modify and control her bodily needs in order to minimise time spent. She says: 'If there was a line, I wouldn't go number 2 [defecate] just because I would be self-conscious about taking too long.'

There is a huge amount of FASE surrounding defecation in women's public toilets. Many of my interviewees admitted to never defecating in public. This is part and parcel of maintaining *homo clausus* boundaries generally and is implicated in each rule of the TIO for users of women's public toilets. Here, since *homo clausus* feminine sex-gender is already caught in a double standard of irrationality and pristine materiality, users of women's public toilets experience much unrest when it comes to time spent in defecation, as I explore further below.

Beyond the largely accepted practice of queuing for a cubicle (itself an act of bodily management and control), users of women's public toilets minimise the use of their bodies in time and space by controlling what their bodies can or cannot come into contact with and by following ingrained habits when choosing and using cubicles. While in a cubicle, many users try to minimise their bodily contact with the space. This is most readily evident in the hetero feminine practice of hovering. Many users hover over the toilet seat, not allowing their bodies to come into contact with the apparatus *they are in the space to use*. Lucy, a twenty-three-year-old bisexual woman, like many of the users I spoke to, directly relates this practice to her early *homo clausus* socialising experiences when speaking about her mother and childhood:

> I never sit on toilet seats. When I was little my mom was always super paranoid about that, she was always like 'don't sit on public toilet seats!' It is just such an ingrained thing; it's just what you do.

Part of this ingrained behaviour limits what users can do with their bodies; it binds them to a particular embodiment of hetero femininity. Consider that it becomes especially difficult to hover and do anything else when you have to hold your coat or purse (also keeping them from touching anything 'dirty') or the cubicle door shut, for example. (This may explain why so many mobile phones meet their demise in toilet bowls.) One queer woman I interviewed described this as 'bathroom yoga', making light of an annoying and difficult reality. A reality which renders defecating nearly impossible as it is exceptionally difficult to hover over a public toilet seat and defecate, let alone defecate quietly; which is a mandate of the third dimension of the TIO and a practice which requires that users exercise intense control over the orifices of their bodies. Regarding habitual use, even when users don't have to queue for a cubicle, when they have the luxury of choosing, users tend to not fully use the space available. Instead, they simply replicate their individual patterns of use, keeping everything about the process direct, minimised, and controlled.

Elizabeth, twenty-four, queer woman: I always use the first cubicle if
I can.

Lana, thirty, queer woman: I usually go to the far end of a bathroom
and use the last stall.

The tendency to replicate individual patterns is also a way to minimise
the time spent in these spaces because, generally, engaging in a pattern
that doesn't require much conscious effort or decision making is quick-
er than looking around and choosing a cubicle based on the particular
space.

Similarly, users of men's spaces engage in nearly the same behaviour
but with an even more rational basis, because, typically, men's spaces
include urinals that are visible to anyone moving through the space;
thus users are required to manage their bodies under the potential gaze
of other men. This means users of men's toilets have to be even more
conscious (than users of women's toilets, since the queuing system and
personal patterns tend to direct women) of where they place their bodies in space; being careful to minimise their chance of coming into
contact with other bodies.

By turning to the survey data, we can conclude that there is a clear
order to the way users choose urinals based on other masculine bodies
in the space. Specifically, men always try to distance themselves as
much as physically possible from other men at the urinal. While this
may seem obvious, it is important to highlight that this act of choosing a
urinal, of distancing one's body, is an utterly social act, one which does
not happen independently or individually, rather, it always requires an
active knowledge and conscious awareness of where other bodies are in

Figure 5.1.

the space and how they correspond to the unwritten[19] codes of 'proper' urinal usage. According to my survey data, in a men's public toilet with six urinals, where men are occupying the first and last urinals (see figure 5.1), 72 percent of men surveyed chose to use one of the centre two urinals, with only 3 percent choosing a urinal directly next to another man. The remaining 25 percent opted to use a cubicle or simply leave the space altogether. These numbers show just how exacting the TIO is, as it suggests that there would not be this highly defined pattern of use without the pressure to manage comportment in a particular way. The premium placed on maintaining bodily distance and swiftly and directly choosing 'the correct' urinal may help explain some of the hesitation in the proceeding story.

Rick, a twenty-four-year-old straight man, shared this story with me about a time when he didn't minimise his movement and instead hesitated, engaging in additional, irrational movement.

> This was at work; the whole space is pretty cramped. I went to the toilet and there were people at the urinals, two people staggered over the four urinals. So I was not only next to someone, but pretty close physically—one of them was a superior who is very macho and you have to kind of be a frat guy around, so I started to make an entrance to the urinal between them, but it was SO close, SO CLOSE—I'm like my elbows are touching them! It just feels really weird, so I step back, and then I felt even more weird about stepping out of it! So then I stepped back into it—No, no, I started to step back into it and then I was like NO it is too tight! And then I went to use the stall. So I pulled every taboo in the book. I felt like such a loser, all that hesitation. It doesn't really make rational sense, and obviously they knew what I was doing because I was so clumsy about the whole thing, but I didn't want to offend them, by being like, I don't want to pee next to you, you're too close—but also there is this sense that guys pee in public, like that's what guys do, so there are both of those things that, like if I retreated, that it would be these two things of not wanting to, like saying I feel uncomfortable peeing next to you and then this sense of being called out, 'what you can't pee next to another guy?' The whole thing was very uncomfortable!

Here Rick failed to display economy of movement, taking up additional time and space to reach his goal, based upon how his body felt in relation to other bodies near to him. He struggled with how he felt in his body on the one hand and what he rationally knew he was supposed to do on the other hand. He was caught between the need to minimise his contact with other bodies, to minimise the time spent in deciding where to place his body, and the hetero masculine ability to be near other male bodies. Ultimately, for Rick, this display of indecisiveness was embarrassing and somehow, he felt, revealed something about his 'nature'—about his ability to 'naturally' 'do what guys do'.

Similarly, Zevi, a twenty-three-year-old queer man, speaks about the awareness he has of his body when entering the space and how he adapts his movement from a looser, more open, 'queer' style of walking to more of a 'straight', economic, and restricted one. He says, 'Like for me, I think about whether like some queeny fag is going to swish on into the restroom or whether, well, I'll like, try and move very rigidly and unnaturally.' In Zevi's example, movement, which is expressive of and a valued part of his identity, is not valued—it is queer, excessive, superfluous of what is rationally necessary—so instead he consciously changes his style of embodiment to match that of the *homo clausus* hetero masculine norm. While Zevi doesn't personally 'identify' with this norm, within these spaces he 'naturally' (without difficulty) adopts the material-discursive practices expected of him in order to remain legible and non-threatening. In both Rick and Zevi's examples movement which is not directly goal oriented is understood to connote some greater truth about that person's identity, showing how *homo clausus* dis-embodiment restricts sensory-embodiment to what is rational, orderly, and consciously managed.

Minimising one's movement in the time-space of public toilets requires a set of material-discursive practices that seeks to maintain the imagined borders of one's body. For users of both men's and women's public toilets this management is rarely questioned and regularly engaged. Such patterns of use, according to Ahmed, are directive and directed. She says, 'Lines are both created by being followed and are followed by being created. The lines that direct us, as lines of thought as well as lines of motion, are in this way performative: they depend on the repetition of norms and conventions, of routes and paths taken, but they are also created as an effect of this repetition.'[20] It is through these self-generating lines that deviations can easily be spotted. By limiting the time spent and minimising the use of one's body in space via social and personal patterns—repetitive lines of use—one can limit one's exposure to the potentially threatening space of open, excreting bodies, an abject reality that fundamentally stands in opposition to *homo clausus* subjectivity. In order to avoid the abject—the breaking down of bodily borders—the *homo clausus* subject maintains a precarious relationship to FASE, which propels one to follow the rules of the intra-action order. An important facet of that management concerns how, where, and when the senses can be used.

RULE TWO: MIND YOUR EYES (RATIONAL)

According to *homo clausus* sensorial individuation (i.e., the division and counting of the senses which allows sensory happenings and observations to become the domain of the rational mind), seeing is of crucial importance. Not only is sight believed to be the sense most closely related to the mind (if it is experienced as such, it is because of this process of sensorial individuation), it is vital in the management of the abject. As *homo clausus* individuals use their sight to do a huge amount of their sensing and monitoring—instead of touching we merely look with the 'mind's eye' at something, and not physically coming into contact with it helps keep the imagined borders of the body safe and stable—the gaze within public toilets is intensely and expertly managed in the effort to maintain distance between bodies and hetero sex-genders and to firm up bodily boundaries. This is a process of structuring and maintaining desire as much as it is about bodily distance. Like in the previous rule, this rule operates in equal yet seemingly opposite ways in the two spaces, again enabling material-discursive practices to solidify sex-gender difference. For users of men's spaces, the role of sight is one of expected self-censorship, regardless of sex-gender-sexuality, because the male gaze is typically an objectifying gaze and there is seemingly nothing to objectify (no women) in the hetero masculine spaces of men's public toilets. For users of women's spaces—i.e., those who are typically objectified by men—theirs is 'an inspecting gaze, a gaze which each individual under its weight will end by interiorising to the point that he is his own overseer, each individual thus exercising this surveillance over, and against himself.'[21] Users of men's public toilets refuse to look at one another in an effort to avoid objectification (an expression of desire) while users of women's public toilets look at one another to ensure hetero femininity is being upheld and not exposed to the opening of sexual objectification by any women. For in both cases, the sense of sight is used to manage the imagined borders of *homo clausus* bodies; masculine users remain closed by not looking and feminine users look in order to remain closed.

Users of men's spaces, when speaking about their behaviour generally, had the following to say in relation to the use of eye contact:

> Jared, twenty-eight-year-old, straight man: You generally don't look at people or talk to them.
> Justin, thirty-three-year-old, queer trans man: In the men's room there is no eye contact and little to no acknowledgement.
> Jason, twenty-eight-year-old, queer trans man: My concern would be more about just avoiding eye contact with everyone.

The material-discursive practice of avoiding eye contact means that men's public toilets are generally experienced as heteronormative spaces regardless of sex-gender-sexuality. This is evidenced in the pervasive understanding that those who do not keep this practice are inherently non-normative and deviant. For example:

> Emit, twenty-seven-year-old, queer trans man: There is no eye contact, and if there is, you're gay.
>
> Ford, twenty-four-year-old, queer trans man: Men don't pay attention to what anyone else is doing, and if they do, they're gay.

Some who imagine they may be on the receiving end of a glance from another masculine user experience this as an inherent threat to their own hetero masculinity and have a strong desire to discipline the 'looker' through masculine (often 'penetrating') violence. For example, Ash, a thirty-two-year-old straight man, says, 'If I saw someone looking at my dick at a urinal I would piss in their eye.'

Those who are non-normatively sexed-gendered clearly represent an inherent threat to closed masculine bodies. Men who desire other men represent an openness and fluidity between bodies which is disturbing to *homo clausus* identity. *Homo clausus* bodies are so intensely managed that even a simple glance is understood as suspect and threatening. Here sight is another condensation point of FASE. As Edelman states, 'The law of the men's room decrees that men's dicks be available for public contemplation at the urinal precisely to allow a correlative mandate: that such contemplation must never take place.'[22]

Furthermore, there is the need to strictly deny the intense erotic potential inherent in these spaces. According to Jeyasingham,

> Legitimate use of the urinal, use that asserts its functionality, depends on *looking straight* (ahead or down), and that incognisance to the looks of others in the room (whether for all intents and purposes waiting for stalls to become vacant, or washing their hands methodically and scrupulously), to the extent that it refuses to acknowledge the pissing male figure's erotic potential, multiples it.[23]

Other bodies, other 'dicks' in the space become the proverbial elephant in the room. This reality has the potential to create an intensely charged space which men are, within the TIO, only allowed to deny. While this was implied in many of my interviews, very few men spoke about it openly. Steve and Erik, both twenty-four-year-old straight men, while speaking about their experiences at the urinal had the following to say:

> Steve: I feel like we're all trying to avoid trying to look at each other's dicks so we're not looking at anything, so yeah I'm not trying to,

or trying not to make eye contact with other people in the bathroom, definitely when I'm peeing I'm trying not to look at anything except for the wall in front of me or my stream of urine.
Erik: Like there is, there is the thing, like, about, other people seeing your penis, or something like that. It's like, wouldn't it be
weird if this happened and like, I don't know, I don't know, it's
just like, try to avoid this potential thing from happening . . . the
awkwardness comes from other people seeing your private parts.

From a slightly different angle, Kel, a thirty-three-year-old straight
man, who prefers to avoid the urinals altogether and urinate instead in a
cubicle with the door left ajar, speaks about his internalisation of the
pressure to hide his body from other's eyes in a moral or religious sense,
ultimately connecting it to desire, love. Overall, there is a strong conno-
tation of sex and sexuality in his explanation:

> My biggest concern is visual privacy. I try to live a pure life and in
> some way having my wiener [penis] out in front of other guys to see
> feels impure—I think there must be some religious or moral under-
> tone to this feeling. It's like being naked in public, but not really, it
> just feels inappropriate. I think you should only be naked and show
> your body to your partner or loved one so it isn't something I like to
> do in public, even in the bathroom. I find that trying to maintain a
> sense of purity is just really hard to do in the bathroom.

Despite the fact that men are very aware that they should never look at
each other's dicks in a public toilet, Kel uses this potential as a justifica-
tion for his behaviour and related discomfort in public toilets. Ultimately, these experiences highlight the power given to *homo clausus* sight
and the nature of the masculine gaze to objectify and sexualise, as well
as how men work to manage bodily FASE.

In women's public toilets, rather than objectifying one another, the
gaze is employed for hetero feminine policing. As Bartky highlights,
'Normative femininity is coming more and more to be centred on woman's body—not its duties and obligations or even its capacity to bear
children, but its sexuality, more precisely, its presumed heterosexuality
and its appearance.'[24] As the TIO of women's public toilets are heteronormativising, all bodies that enter are expected to at least conform to
homo clausus hetero feminine behaviour and appearance. Put simply,
women look at themselves and other women in public toilets to make
sure they appear to be hetero women and thus non-threatening, legitimate users of the space. This is why, for instance, users who know or
fear they do not immediately appear hetero feminine adjust their bodies, voices, and clothing in order to show other users that they are

indeed in the correct space, though this is often not enough, as the visual, enacted via hetero selective attention, takes primacy in *homo clausus* ways of being. Much of this looking occurs through the space of mirrors. The mirror in a public toilet is an important tool for asserting hetero femininity, which is why many users of women's public toilets who do not identify with it feel uncomfortable using the mirror. Like washing one's hands, the use of the mirror seems to be a 'cleansing' ritual that, after the dirty, destabilising use of the cubicle, re-establishes a user's femininity. The mirror space is the only place within women's public toilets where users feel they are permitted to take their time. The use of the mirror is an important performance and public embodiment of conventional femininity and many users speak about using the mirror as something that is normal, natural, and necessary:

> Cece, twenty, heterosexual woman: Retouching your make-up is a big one . . . there is a need [to use the mirror], especially if you go to the toilet in the middle of the day and you need to look at yourself or if you're having a night out and go to the toilet, I think there is a need to look in the mirror because a lot of girls care about how they look, makeup wise, in public.
>
> Rachel, thirty-seven, lesbian woman: When I use the toilet I always feel compelled to check myself in the mirror; it is very much a part of being female, I'm not sure why I do it but I know it is what I'm supposed to do.
>
> Emit, twenty-four, queer transguy: Yeah in women's bathrooms everybody cares about their looks and appearances and is checking themselves out.

Many trans, lesbian, and queer users I interviewed, who use women's public toilets daily, have a complicated relationship with the mirror. They speak of it as an uncomfortable pressure, a practice that makes them uneasy.

> Elizabeth, twenty-four, queer woman: There is definitely a pressure for vanity for women. . . . I'll walk out of the bathroom and think maybe I should've checked the mirror.
>
> Lana, thirty, queer woman: Yeah, I do look in the mirror and I do feel self-conscious about looking in the mirror. . . . I don't like when people catch me looking in the mirror.

Frankie, who is twenty-five and queer, takes this discomfort one step further and considers this practice in relation to her gender and sexual identity:

> I've been thinking about how I never look in the mirror when I wash my hands but everyone else does, kind of for a while to fix things, I've never done that my whole life. I'll look in a mirror, but I won't stand there and preen and I think maybe, largely it is part of my sexual-gender identity. I know I'm a woman, but I feel more like, not more like but just that I don't fit in with women and so maybe that pardons me. It's not that I'm better than this, it's just that I don't do it and I feel like I'm not supposed to do that or don't need to do that.

While aware of the pressure for hetero feminine displays of vanity which overtly satisfy the policing gaze, Frankie feels that she isn't supposed to use the mirror because she isn't heterosexual and conventionally feminine. She still feels the pressure to display the social identity of conventional femininity by using the mirror, even though she knows she already undermines and threatens it through her unconventional queer identity.

The use of the mirror and an inspecting gaze generally are also used for overt, non-self-imposed policing of hetero femininity where users are, for example, able to call upon state police officers and/or security personnel to further inspect users' bodies who are believed to be the threat to femininity incarnate. Throughout my research I heard many stories of users' bodies being visually policed, supporting the *homo clausus* notion that to see something is to know it. Sam, who is a twenty-four-year-old queer transguy, told me about a policing experience when he and his girlfriend went into a women's public toilet together. He explains:

> This was before I began my transition [to masculinity] and I was identifying as genderqueer. Two women in the bathroom got the police and told them there was a man in the women's room. Then two male police officers came into the bathroom to seek me out. My girlfriend vouched for me, said I was female, that we were in the right place. It makes me wonder why was I so scary to those women? I mean should I have to show my ID to be able to use the toilet? This experience was a point of anxiety for me about my body and public life.

Beck, who is fifty years old and a butch lesbian who uses women's public toilets 'with complications every time', shared this story with me about a toilet attendant:

> She [toilet attendant] actually followed me out of the bathroom and across the street, into Starbucks because she still couldn't tell [if I was a woman] and wanted to make sure [of my sex-gender] and she waited until I got my coffee and she was pretending to clean up

there! And I said to her, do you want to talk to me? She just got all
flabbergasted and walked away. She didn't want to talk to me, she
just wanted to look at me, and probably see if she could get me
arrested.

Billie, a twenty-three-year-old genderqueer person, spoke to me about
hir regular policing experiences:

I've been told by women that I'm in the wrong restroom so many
times that I've come to expect it every few times I use one. One of
my most vivid memories of being gender-policed was in an airport
when I was 18 or 19. I was about to go into the lady's room when a
woman walking by shouted at me: 'That's the women's room!' I always feel sort of bad and embarrassed for people when they make
assumptions about me and I have to correct them.

In these stories, users saw an individual who didn't *look* conventionally feminine and without speaking to that individual, or even looking a
bit more closely, immediately felt threatened or, in Billie's case, assumed the person they saw didn't know what they were doing. In Sam's
story, the policers, sure of their judgement and assuming the mere
presence of a man in a feminine space must mean trouble, sought
enforcers of the law. The irony of inviting two 'powerful' men into the
space to further investigate the possible aberrational presence of a man
is both hypocritical and unsurprising. Here conventional femininity
relies directly on patriarchy to reinforce *homo clausus* gender boundaries and manage what is believed to be a threatening situation. Beck and
Billie's stories also expose how closely heteronormative binaries are
bound to appearances and how readily *homo clausus* individuals rely on
selective visual attention, via sensorial individuation, to make sense of a
situation. Despite the fact that each of these individuals had 'female'
bodies (e.g., breasts) at the times of their stories, their visible gender
expression was not overtly feminine enough to keep them from being
policed.

The gaze is particularly important here because it takes precedence
in determining one's right to be in the space. All of the individuals
whom I spoke with, who are policed in their daily lives, expressed frustration with the refusal of their policers (those who challenge another's
right to be in the space) to speak to them, to simply *listen* to their
feminine voices before making a judgement about their gender. Instead, users judge who is feminine 'enough' through the appearance of
hetero feminineness; despite the fact that users of women's public toilets, who are regularly policed in these spaces, already modify their
behaviour in accordance with this critical gaze (e.g., by removing layers

of clothing or sticking out their chests), the policing continues. This point highlights the instability, fear, and anxiety connected to feminine embodiment and sexuality, and how it feeds the immense yet dispersed pressure users feel to conform their immediate appearance and style of looking (in both senses) to match what is expected by the policing feminine gaze. Furthermore, while the surveillance and policing of other bodies in women's public toilets is justified in the name of safety, it seems obvious here that the threat to users in these spaces is rarely a threat of crime or violence (though it does happen, just not nearly as often as queer users are policed). Instead the apparent threat is to the reproduction of hetero femininity. As feminist academic and author Sally Munt (interpreting Eve Sedgwick) explains, '[The butch] instigates female homosexual panic amongst the women, a violent reaction which betrays the disturbing belief that sexuality is the solvent of stable identities.'[25] She goes on to explain, 'I am painfully aware that being challenged about one's sex is not usually the issue; my body is read "correctly" as female, but my gender causes the problem, hence the question "Are you a man or a woman?" is a displacement of the unutterable "Are you a lesbian?"'[26] Munt, writing from her own experience, powerfully shows the entanglement of sex-gender-sexuality both in experience and in discourse. Through onto-epistemological, material-discursive practices, her example nicely highlights the intra-acted nature of heteronormativity. In the preceding examples then, it would be naive to merely dismiss the intense disciplining of the body within these spaces as a by-product of a necessary way to keep the space safe. Instead, the TIO works mainly to keep users' bodies in line with *homo clausus* heteronormativity by continuously fuelling the threat within the space. However, the bodily acts which require the most intense hetero feminizing, in order to stabilise *homo clausus* subjectivity, involve those that prompt most individuals to use the space in the first place, excretion.

RULE THREE: MANAGE YOUR BOUNDARIES (INDIVIDUALISATION)

The third dimension of the TIO of public toilets is concerned with maintaining social and rational individuality while engaged in acts that inherently transgress bodily boundaries. Transgression of this sort is intensely tied to FASE as it is instilled through some of the earliest experiences we have in relation to our embodied-selves. Such bodily experiences, where something from the inside of one's body makes its way to the outside, are typically experienced as abject and non-personal, and learned to be managed according to feelings of FASE. During bodily excretion the possibility for losing or transgressing *homo clausus* boundaries is very high, but not a given, since the boundaries are largely rationally imposed, not necessarily based on experiential reality. They exist in a state of mentally imposed reflection, which is embodied through material-discursive acts. Bodily boundaries are material-discursive, propping up both socially normative understandings of the 'natural' body and the socially appropriate behaviours in which the body participates, with neither being innate. That is to say, even though bodily excretion always already reveals the unstable and unrealistic nature of the bounded *homo clausus* identity—because it explicitly exposes how the body is not constant, not stable, and not sealed in everyday, mundane ways—one is able to rationally keep the self together through this directly lived threat (bodily excretion) via an emphasis on learned experiences of bodily FASE in relation to the imagined borders of the body.

What's more, instead of a point of convergence, recognition, or bodily openness, acts of excretion in public toilets are normatively construed as a personal aberration that must be managed accordingly. They are acts that are experienced and thus understood as purely singular, despite their utter ubiquity. This, again, is how the bodily is not merely discursively bounded but how bodies in space are material-discursive. Like *homo clausus* generally and the other two rules of the TIO, this rule is never enacted from a point of pure individuality, but rather exists socially, entangled with other bodies in time and space; it is not individual but rather, works to make one *feel* individual. Therefore the material-discursive practices and experiences that go into managing one's boundaries are reliant upon the socially situated nature of the TIO for both men's and women's public toilets. Collectively held and deeply ingrained assumptions of *homo clausus* bodies are bound by binary gender in public toilets and serve as the basis of socially appropriate, individual bodily conduct and appearance. For men's and women's public toilet spaces, users are concerned not only with maintaining the boundaries of their own bodies, but also in making sure other bodies remain, at least conceptually, closed and sensorially impermeable. Similarly, but observed from a different perspective, Moore and Breeze research how the spatial location of public toilets disrupt social boundaries, insofar as they are not 'regular social spaces' and 'might strike us as beyond social management, control and order.'[27] My exploration here takes a similar approach to boundaries but happens at the scale of the body, showing how when we understand social life from sensory-embodied experience, we can access the extent to which 'social management, control and order' continue to organise our behaviour and support dis-embodied identities.

In women's public toilets there is an interrelated sensory-embodied awareness that permeates the space, fosters anxiety, and encourages the overt management and policing of bodies. This general 'sense' of the space is supported by the first two rules of the TIO which ensures that users of women's public toilets minimise their bodily use of time and space and adopt an inspecting gaze which seeks to maintain hetero femininity. These two rules are operational only because people actively demarcate what is and is not one's body; a process of weeding out sameness and difference which identifies what is an abject other. As *homo clausus* subjectivity is built on an impossible universal bodily likeness, enacted through the division of bodies along hetero gendered lines—a process which turns material differences into a failure of sameness through exaggerated sex-gender difference via discursive practices—awareness, anxiety, and policing in women's public toilets seeks to maintain the hetero gendered division material discursively by identifying and expelling those bodies which do not fit neatly on either side of the gender binary; those bodies which are considered abject. Furthermore, as Bartky explains, 'The disciplinary power that is increasingly charged with the production of a properly embodied femininity is dispersed and anonymous; there are no individuals formally empowered to wield it; it is . . . invested in everyone and in no one in particular.'[28] Hetero femininity, since it is both performative and required in public toilets, can easily be threatened by those bodies which do not properly embody it and, as women's public toilets are experienced in mundane daily excretory usage as individual, yet collectively managed, all bodies have the potential to be threatening. This is because each body, each individual, is expected to entirely discipline and manage themselves without interfering with other individually managed bodies; a compulsion that attempts to entirely obscure the social, entangled nature of the action. When bodies are controlled in such a way—where each body is radically alone in its experience—those who fail to maintain the standards of such management and control leak out into the space in a disturbing and apparent transgression of the individual boundaries of the body. This management of bodies operates through one's general spatial awareness of where bodies are in the space and what acts they're engaged in, which fosters and furthers bodily anxiety regarding the transgression of sensory boundaries during excretion. I will deal with these in turn.

First, a general sense of awareness of bodies in space is required in order to determine the intensity and style of boundary management required for one to engage in at any given time. A crucial aspect of

managing one's boundaries is sensing other bodies in the space. For
example, users I interviewed, while speaking about their use of the
space generally, had the following to say:

> Alice, twenty-six, queer: I think I'm quite aware of other people in
> the space. I don't know even why or how, just when I go in, I'm
> not even really paying attention, but I immediately scan the place
> and you know where other people are in the toilets and what
> they're doing.
> Cece, twenty, heterosexual woman: Well, I think I'm pretty aware of
> what everyone is doing in there. . . . If I am in a place where I
> know people, I can usually tell who is peeing where, just by the
> way they are in the cubicle.

The first two dimensions of the TIO create spaces which are primed for
self/other policing, while a general sense of awareness capitalises on this
dynamic and helps maintain the feeling of needing to actively manage
one's boundaries. This is evident in how users speak about always knowing where other people are in the space and what they are doing. This
knowing or general spatial awareness is a step towards overt policing; it
is a less intense yet more pervasive management of body-selves which
provides an understanding or embodied knowledge for knowing to what
extent one has to manage their boundaries—if the space is empty or if
the space is very busy (and noisy) the pressure to manage bodily boundaries is extremely reduced as the imperative to be responsible (to
other users in the space) for what one's body produces, for the transgression of bodily boundaries, is severely lessened and often nearly non-
existent.

This awareness or material-discursive process of knowing, rather
than simply rational, is a sensory-embodied knowledge which is seldom
understood as such. The low-grade anxious awareness that permeates
women's public toilets is an example of sensorial individuation in action.
It is an awareness which has been individually conditioned to work
without overt mental attention but satisfies a rational anxiety which is
not biologically or practically necessary for using the space. Rather,
one's senses are employed and deployed to form an awareness of bodies
in space which acts to mentally form an understanding of one's own
body in space through rational sensing of difference and distance according to reflective *homo clausus* understandings of self-body subjectivity. The awareness seeks to maintain and preserve hetero femininity
at the individual level via FASE. It is not a sensing of becoming but
instead of judgement, of anxiety.

This process of being aware of, looking at, smelling, and listening to
other bodies is a prerequisite of policing. Cultivating this sensory-indi-

viduated awareness necessarily requires users to keep their own bodies in line with the TIO through psychically projecting the self outward (instead of cultivating an embodied awareness) and back onto the borders (outer surfaces) of the body. It creates an anxiety which keeps users hyperaware of their bodies, emphasising the need to rationally maintain the imagined borders of one's self-body. Users spoke about mundane usage of public toilets in the following ways:

> Natalie, twenty-four, queer: For me, the experience is anxiety provoking. I always feel like I'm going to be judged somehow on what I'm doing. By even, like, the sound of the way I use the toilet paper.
>
> Cece, twenty, heterosexual woman: I definitely don't think they're relaxed spaces.
>
> Frankie, twenty-five, queer: I think everyone, including myself does have anxiety about bathrooms, it is the norm.

The material-discursive practice of building spatial awareness of bodies in space and of what they're doing is preservationist and anxiety producing. It simultaneously strengthens the sense of one's bodily boundaries while exposing their weakness and openness to interruption. It is through this individualisation, this solitary *individual* experience, that the experience of atomisation and the ongoing anxiety that one may accidently transgress one's bodily borders is perpetuated.

Second, bodies engaged in acts of excretion represent interruptions and fissures in the presumed cohesiveness of bodily boundaries. They reveal the unstable nature of bodies. Such realities, while inherently part and parcel of public toilet spaces, are felt to be disturbing to all those present, an individual deviation. Users feel that their bodies, when engaged in excretion, are beyond the limits of their boundaries and thus out of control and invasive. For example, when speaking about excretory acts, users explained:

> Alice, twenty-six, queer: There is always this thing, especially with women I think, that you're always very careful not to disturb anyone.
>
> Miriam, twenty-five, queer woman: I feel guilty that I'm disrupting other people's bathroom time. It's not like anyone would ever be, or I've never experienced, after tooting [passing gas] or shitting in a public restroom with other people around, like someone giving me a dirty look or asking me, why did you do it? That has never happened, but I think that I imagine that is what they're thinking.

These feelings are based in rationality, not in experiential reality. While users I interviewed have never experienced any overt social ridicule in

relation to excretory acts, their inherent transgression of bodily boundaries is experienced as abject and continues to propel self-enforced
FASE. This transgression is experienced as such because of the sensory.
While users' bodies in excretion are not normally seen by other bodies
in excretion (or generally in women's public toilets), they do come into
contact through other sensory modes, which are judged and mediated
according to feelings of FASE. As a result, an uneasy, tentative atmosphere exists in the space. It produces an anxiety concerned with maintaining hetero feminine, *homo clausus* behaviour. Despite intense management of bodily boundaries, one cannot always choose what one
senses, and this reveals a fundamental reality about *homo clausus* individuality: no matter what, it is never individual. The spatial awareness
(explicated above) produces a tension which is both a manifestation of
the expectation to manage and maintain the boundaries of one's body
and also a way of perpetuating the threat to them. The awareness and
tension also work through self-other transgression and sensory infiltra-
tion, which are highly threatening to hetero feminine *homo clausus*
identity.

The sensory space users are most concerned with transgressing is
sound, often coughing or rustling toilet paper or clothing to mask the
bodily sounds they can't avoid making. There is a code of silence in
women's public toilets that should only be broken by conversation at
the sinks or mirror, not by a body within a cubicle. The taboo on the
apparent transgression of bodily boundaries tries to keep hetero femininity stable, while exposing its reliance on the body. Keeping in mind
that the muted feminine body is an ideal of patriarchal culture, the
transgression of sensory barriers when expelling bodily waste is highly
problematic and threatening to *homo clausus* hetero femininity.[29]
When users do defecate, a natural bodily function that many users
admit to never doing in public, the pressure is immense, as the risk for
shame and exposure of the fragility of the 'stable' body is high. When
defecating one often has to pass gas, or fart. Emitting sound even in the
confines of a private cubicle, within a space seemingly built precisely
for the act, is usually experienced as highly embarrassing and de-femi-
nizing. As evidenced in the following quotes, these universal bodily
functions are highly problematic for users:

> Miriam, twenty-five, queer woman: When I'm using a public bath-
> room, I feel uncomfortable, I feel guilty if I fart or poop.
> Lana, thirty, queer woman: You try not to make a lot of noise in
> there, always, like if you're having a noisy poop or any kind of
> poop, you flush or cough to cover the sound, or you don't move or
> make a sound and just wait until you're alone.

Frankie, twenty-five, queer: If I have to poop and I'm at school . . . it makes me really nervous and I'll time it for when someone else flushes or makes some sound, I don't know how I do it, but it's very intentional.

Natalie, twenty-four, queer: In a more intimate setting of two or three cubicles and you need to poop and you know the other cubicles are filled, I feel like I need to wait and be really quiet and wait 'til everyone leaves to poop.

Kelly, twenty-eight, heterosexual woman: I had a friend who would wrap her arm in toilet paper and catch the poop as it came out to keep it from making that horrible plopping sound . . . I thought that was so clever!

Additionally, when users do defecate in public there is also a fear that they may not be able to distance themselves from the physical evidence of their taboo transgression. Cece, a twenty-year-old heterosexual woman, speaks about how she copes with this fear in public toilets:

> Well, I usually avoid doing it [defecating], I need to have done it there before just to see that the flush works and that they have a plunger just in case. I wouldn't easily go, especially if there is a queue. I'd rather go to a disabled toilet where people don't normally go.

The use of the disabled toilet in this situation is acutely understood by Munt who, again, draws on her own experience: 'Using this toilet is inflected by shame. I am . . . not "worthily" disabled, but certainly afflicted. It is at once a perfect, and anachronistic designation, the same positioning simultaneously dis- and en-abling.'[30] Cece goes on to explain:

> The thing is if there is a toilet which had issues flushing in Bombay [India], I wouldn't remotely think about not using it because it isn't a big deal, it is just here it isn't acceptable behaviour for girls in Britain.

Cece, who is originally from Bombay, India, has realised how her feminine identity and related behaviour has shifted since moving to England for university. She has consciously adopted *homo clausus* ways of being as a young adult in order to properly embody the hetero femininity expected of her in England. Part of that includes learning to embody FASE regarding her bodily functions generally and particularly around any visual evidence of such functions. On several occasions throughout our interview she explained that there were many aspects of toilet use in England that 'were strange at first' but to which she has gotten accustomed and now views as normal and necessary.

Taken together, the immediately preceding examples show the various concerns and practices that go into managing the threat faced by bounded bodies in women's public toilets and how those threats are both mitigated and sustained through awareness and anxiety surrounding bodily transgression. In this dimension of the TIO we can further see how all bodies in women's public toilets are expected to follow specific regimes of conventional feminine appearance and performance according to *homo clausus* bodily ideals.

Men's public toilet spaces operate in accordance with this rule in nearly the same ways as women's public toilets: awareness of bodies in space and the fostering of a general bodily anxiety. The experience for users of men's spaces, though, is more concerned with masculine heterosexuality, whereas users of women's spaces are generally concerned with keeping their bodies in line with hetero femininity (where heterosexuality is implied according to sex-gender-sexuality). The emphasis is less on gender expression for men because of the nature of phallocentrism, and more on the expression of desire. The awareness and anxiety surrounding the shoring of masculine bodily boundaries is expressed in terms of keeping straight the presumed heterosexuality of excreting bodies. Users, while speaking generally about usage of space, had the following to say:

> Erik, twenty-four, straight man: I just want to go in do my thing and get out of there as fast as possible. . . . So yeah, if I have to urinate, I'll be looking to see if there are dividers between the urinals and if there aren't, how many people are there, what is the likelihood that someone is going to come in, or whatever and then I might use a stall and I try not to touch anything. . . . I definitely wouldn't start a conversation.

> Emit, twenty-seven, queer trans man: In men's bathrooms, if you're social, you're gay. There is no eye contact, and if there is you're gay, and it is sort of like these unwritten rules—you're in there to do a job and get out. . . . So, you see women going to the bathroom together, you don't see men going to the bathroom together, if they do, they're gay. There are all of these things and they're somehow linked to an expression of your sexuality.

> Robby, 32, straight man: If there is a line of urinals and you have the option of choosing, you cannot choose any urinal next to someone. It is just really gay.

These statements expose the underlying thought processes and feelings surrounding use of the space generally (as related to *homo clausus* onto-epistemology) and the material-discursive processes mandated by the first two rules of the TIO. The heteronormative goal-oriented nature of the material-discursive processes engaged in by users of these spaces is highlighted in these quotes; they point to how anything beyond the absolute necessary movement and usage of the space is anxiety provoking, because it is read as gay. The awareness and anxiety around being read as gay is experienced by heterosexual, gay, and queer users alike. Here sexuality is bound up with the level of one's masculinity, and the imperative in these spaces to be bounded and closed also means to be categorically masculine (another example of sex-gender-sexuality). Leaky, unstable, out-of-control bodies are historically negatively associated with the feminine (which they must work at managing according to *homo clausus* ideals) through patriarchal heteronormative constructions of gender. It seems the nature of excretion in public inherently threatens hetero masculine *homo clausus* subjectivity because it reveals the undeniably open, fluid nature of masculine bodies too. Thus there is an immense amount of bodily awareness that goes into keeping masculine bodies rigidly directed and closed.

Admission of masculine bodily openness is fundamentally inimical to the sealed, bounded nature of *homo clausus*, which often means that users of men's public toilets have trouble with literal acts of bodily openness. Since *homo clausus* identity requires that individuals are bounded and rational, not open in sensory-embodied becoming, excretion is difficult for many. The experience of stage fright—being unable to pee—at the urinal makes for a particularly revealing example of the pressure to remain closed and individually bounded. Steve, a twenty-four-year old straight man, speaks candidly about this experience:

> I have stage fright. That's what they call it and it is a weird thing because I'm totally conscious of it and it's not, there is no emotional sensation or anything, it is literately like I am completely comfortable right now, there is a dude standing next to me, I'm relaxed but I just cannot pee because there is a dude next to me; I cannot explain it, it is like some magic of science, it is this mind boggling thing. So if I'm in a busy crowded bathroom and I see that there is a stall open, I'll go for the stall most of the time because I know that I'll pee and there is no pressure and the only time that I do start to feel weird about it [stage fright] is when I'm sort of like, when I do go to a urinal and it is crowded and I have stage fright and I'm just like OKAY I'm just going to pretend like I'm peeing now, I hope nobody can tell. . . . So, it's more like, it's like I don't know, my mind and body doesn't want it [urination] to happen, because I can't help it and then I have stage

fright and I do become more embarrassed. . . . There was definitely a time when the idea of stage fright, I was not comfortable with that and thought that, not that I had some problem, but that I was like some kind of wuss or something.

Steve's experience of stage fright shows that despite his status as a straight man (who is not interested in having sex with other men), he still has trouble performing this private bodily function in the presence of other men because bodily openness is understood as gay, deviant, by *homo clausus* subjectivity. Additionally, since all action in men's public toilets is expected to be directed and purposeful, standing at a urinal with your penis out, but not being able to perform the 'simple task' of urinating can be highly problematic. While it is something Steve has accepted as part of his life, he is unable to stop it because the embodied pressure remains. To cope, he disconnects himself, removes his agency from the situation and blames his 'mind and body'. Shilling notes that individuals 'frequently experience their bodies in a number of ways as being beyond control'.[31] Therefore, in order to maintain his straight masculine status and preserve the *homo clausus* borders of his body he either hides in a cubicle or pretends he is able to accomplish the task at hand in order to enact some control.

To further understand how intensely men have to monitor and control their bodies while enacting this and the previous rules, it is useful to look at ways some men remove themselves from the TIO by limiting their ability to come into contact with other open, excreting bodies through the use of a cubicle or by leaving the space altogether. According to my survey data, men's public toilets which had three of the six urinals occupied, surveyed in two different spatial configurations (see figures 5.2 and 5.3), showed that 12 percent of men surveyed would turn around and leave the toilet space if they were confronted with these arrangements in a public toilet, completely removing themselves from the TIO and not having to publicly reveal the openness of their bodies or subjecting themselves to the possibility of being read as gay. The presence of 'too many' masculine bodies in these spatial configurations renders some users unable to carry out their hetero masculine duty mandated by the TIO, and they would rather ignore their bodily needs than risk error in performing the intra-action order. Similar awareness and anxiety surrounding the opening of the closed, bounded hetero masculine body is evidenced in the use of the cubicle. The same two spatial configurations (figures 5.2 and 5.3) nearly doubled the use of the cubicle when compared to when just one fewer body was positioned at the urinals (figure 5.1): 43 percent of men surveyed opted to use a cubicle in figure 5.2 and 5.3, while only 23 percent of users surveyed opted to use a cubicle in a room that had just one fewer person (figure 5.1). Taken together, these two sets of data reveal that over 50 percent of men surveyed would rather use a cubicle or leave altogether when just *half* of the urinals are in use. These findings would seem to suggest that when users are required to literally expose the openness of their bodies in public toilet spaces, many prefer to opt out of the intra-action order by either leaving the public toilet space without urinating or defecating, or by entering directly into a cubicle, shutting themselves away from the other bodies in the space. These are actions that carry a potential cost such as physical and social discomfort, loss of opportunity to use the toilet, embarrassment, and extra time spent on finding an alternative.

Additionally, while using a men's public toilet is necessarily an individual activity it does not always begin as such. The taboo is not so much around users physically going to the space together, but instead on any indication that two users within a masculine space may be together in any way. This sort of open sociality is again read as 'gay' and risky by *homo clausus* subjectivity because it overtly reveals an openness between what are expected to be closed hetero masculine bodies in the already threatening spaces of bodily openness. Users, on several occasions, expressed to me how when they go to a public toilet with a friend, any conversation that was happening between them immediately and unthinkingly stops upon entering the door of a public toilet and does not resume until they've exited the space. In this way, there is also a particular social code regarding sound in men's public toilets, as there is in women's public toilets, but one that functions in the opposite way. While silence in women's public toilets operates in order to keep users'

Figure 5.2.

Figure 5.3.

bodies closed and managed according to hetero feminine gender (a silence that can only be broken in polite conversation at the mirrors not by out of control, excreting bodies), in men's public toilets sound is restricted to bodily sounds; any conversation is considered superfluous, suspect, and threatening to *homo clausus* masculinity. Put simply, any social activity is generally prohibited in men's public toilet spaces because it represents an opportunity of and for bodily openness. While social activity, like simple conversation, may not be problematic in other situations, in the context of men's public toilets where bodies are open in irrational fluid processes, everything else needs to be intensely managed in order to keep the imagined borders of the body firmly in place. Despite this understanding, some users, when they do happen to enter the space with another user, feel the need to overtly prove to other users in the space that they are definitely hetero masculine and thus not threatening to other *homo clausus* subjects. While a bit unusual, Steve shares an interesting method he and his friends employ to mitigate the pressure to maintain the imagined borders of hetero masculinity when in the precarious situation of entering the space together. He explains:

> And sometimes we [me and a friend] walk into a public toilet and it is suddenly quiet and we know that anything we say, every other dude in there can hear, so on purpose we'll say something like, 'oh how was the chick you fucked the other night?' ya know just something, ya know, R-rated, X-rated.
> [Is it always sexually charged?]
> Oh no, that was the example I used, um, no I would say they are not necessarily sexually charged, but usually, probably more profane, or related to where we are right then but even in that sense it is

usually like, 'oh did you see that hot chick over there?' which again, yeah, relates back to sex.

Users of various sexual and gender identities all seemingly understand that they are supposed to perform a clear display of heterosexual masculinity regardless of their personal identification. In Steve's example, when his display of hetero masculinity may be unclear, he chooses to do something conspicuous and speak about a blatantly masculine and heterosexual act of 'fucking a chick' in order to announce that he is not a threat to the *homo clausus* hetero masculine space.

In this third and final rule of the triadic intra-action order the material-discursive practices employed by users of both men's and women's public toilets are concerned with acts that firm up bodily boundaries and clearly state that their bodies are in line with *homo clausus* ideals of boundedness and are therefore non-threatening. Bodies which are disruptive to the *homo clausus* style of bodily management in women's spaces are those which undermine hetero feminine gender expression through 'out-of-control' bodily excretion, while in men's spaces those bodies which jeopardise hetero masculine sexuality through acts which point to overt or inherent openness between bodies are most troubling. Through different points of emphasis this dimension works to condense sex-gender-sexuality onto the *homo clausus* body through material-discursive practices that implicate desire with everyday, mundane acts. In both women's and men's spaces expressions of bodily excessivity are considered abject. Here, where bodies overflow their boundaries, the *homo clausus* subject is at its most vulnerable.

FASE AND THE ABJECT: MANAGEMENT AND MAINTENANCE

Throughout this chapter I have posited that the mundane use of both women's and men's public toilets in daily life is rigidly structured by a triadic intra-action order (TIO). The TIO operates via material-discursive practices which seek to maintain the imagined borders of the *homo clausus* dis-embodied subject. The material-discursive practices are heteronormative and work through a condensation of sex-gender-sexuality into two distinct sex-genders. *Homo clausus* dis-embodied subjectivity is generally the same for all bodies, and it is through sex-specific material-discursive practices necessary for maintaining *homo clausus* that bodies coalesce into separate sex-genders. I have worked to show that the material-discursive practices employed and deployed in public toilet spaces are nearly the same for both men's and women's spaces, yet often operating in equal yet seemingly 'opposite' ways, pointing to the fundamental likeness of identity construction for all bodies in the contemporary West. Generally, the emphasis in men's spaces is on the maintenance of hetero sexuality which implicates gender expression, whereas in women's spaces the emphasis is on hetero gender expression which implicates sexuality. Instead of understanding this as two binary oppositions at work (gender/sexuality and male/female) the TIO exposes how close constructions and expression of sex-gender-sexuality are operating on more of a continuum where subtle material-discursive phenomena are construed, via rational processes, into major differences. In both men's and women's public toilets, bodies are kept in line with the TIO through a strong undercurrent of fear, anxiety, shame, and embarrassment (FASE) which act as 'a straightening device' of compulsory *homo clausus* heteronormativity.[32] It requires all bodies to manifest the same style of sex-gendered being (according to the space they use) in order to not appear suspect and threatening. When bodies deviate they are considered abject, not just because of the acts being carried out (i.e., excretion), but because they directly and fundamentally challenge the bounded, stable nature of *homo clausus* being. FASE are therefore crucial in maintaining the underlying abject potential vital for *homo clausus* subjectivity; these feelings are always necessarily present and requiring rational management and dis-embodied attention in order to maintain the feeling of individual stability. The TIO of women's and men's public toilet spaces serve to keep bodies rigidly, rationally managed in order to continue the re-production of the *homo clausus* individual. Public toilet spaces are where bodies are obviously open and shared in their unboundedness; here they represent one of the most apparent yet potentially devastating threats to *homo clausus* ways of being: undeniable openness. Self-bodies openly experienced, expressed, and explored in their unbounded sensory-embodied becoming are antithetical to *homo clausus* subjectivity and represent an opportunity for new ways of knowing, understanding, and experiencing. Thus the rules of the TIO are not stable in themselves but, like *homo clausus*, are stabilised through habitual patterns of use. Thus they can be easily transgressed or completely ignored in ways that radically diminish their power, allowing for fuller sensory-embodied experiences, greater connection with others, and opportunities for becoming-other. The rules of the TIO normally operate through discontinuity of embodied experience but they are not permanent or without challenge, thus they constitute an opening for cohesion through differential ways of being.

NOTES

The opening quote is from Julia Kristeva, *Powers of Horror: An Essay on Abjection* (New York: Columbia University Press, 1982), 4.

1. Judith Butler, 'Performative Acts and Gender Constitution: An Essay in Phenomenology and Feminist Theory', in *Writing on the Body: Female Embodiment and Feminist Theory*, ed. Katie Conboy, Nadia Medina, and Sarah Stanbury (New York: Columbia University Press, 1997), 402, original emphasis.

2. Susan Bordo, *Unbearable Weight: Feminism, Western Culture, and the Body* (Berkeley: University of California Press, 2003), 27.

3. Susan Bordo, 'The Body and the Reproduction of Feminity', in *Writing on the Body: Female Embodiment and Feminist Theory*, ed. Katie Conboy, Nadia Medina, and Sarah Stanbury (New York: Columbia University Press, 1997), 91.

4. Michael Morris and Cate Sandilands, 'Eco Homo?' (Unpublished performance script from Staging Sustainability, 20 April 2011, Toronto: York University), 17.

5. Kristeva, *Powers of Horror*, 4.

6. Morris and Sandilands, 'Eco Homo?', 17.

7. Morris and Sandilands, 'Eco Homo?', 18, original emphasis.

8. Erving Goffman, 'The Interaction Order: American Sociological Association, 1982 Presidential Address', *American Sociological Review*, 48, no. 1 (1983), 2.

9. Erving Goffman, *The Goffman Reader* (Vol. 7), ed. Charles Lemert and Ann Branaman (London: Blackwell, 1997), 205.

10. Goffman, *The Goffman Reader*, 204.

11. See Rawls 1987, Shilling 1997, 1999.

12. Goffman, 'The Interaction Order', 2.

13. Jennifer 8. Lee, 'Ejection of a Woman from a Women's Room Prompts Lawsuit', *New York Times* [online], (City Room) 9 October 2008. Available from: http://cityroom.blogs.nytimes.com/2007/10/09/ejection-of-a-woman-from-a-womens-room-prompts-lawsuit/.

14. Sara Ahmed, *Queer Phenomenology: Orientations, Objects, Others* (Durham, NC: Duke University Press, 2006), 87, original emphasis.

15. Karen Saunders, 'Queer Intercorporeality: Bodily Disruption of Straight Space' (a thesis submitted for the degree of Master of Arts in Gender Studies at the University of Canterbury, Christchurch, Aotearoa/New Zealand, 2008), 127.

16. See Goffman, 1997.

17. Sandra Bartky, 'Foucault, Femininity, and the Modernization of Patriarchal Power', in *Writing on the Body: Female Embodiment and Feminist Theory*, ed. Katie Conboy, Nadia Medina, and Sarah Stanbury (New York: Columbia University Press, 1997), 131.

18. Judith Plaskow, 'Embodiment, Elimination, and the Role of Toilets in Struggles for Social Justice', *Cross Currents*, 58, no. 1 (2008), 52.

19. Though a quick Internet search will show that men have started writing these codes 'down' via games, blogs, videos, and other media showing how men should conduct their bodies in space.

20. Ahmed, *Queer Phenomenology*, 16.

21. Michel Foucault, *Power/Knowledge: Selected Interviews and Other Writings, 1972–1977* (New York: Vintage, 1980), 155.

22. Lee Edelman, 'Men's Room', in *Stud: Architectures of Masculinity*, ed. Joel Sanders (New York: Princeton Architectural, 1996), 153.

23. Dharman Jeyasingham, '"Ladies and Gentlemen": Location, Gender and the Dynamics of Public Sex', in *In a Queer Place: Sexuality and Belonging in British and European Contexts*, ed. Kate Chedgzoy, Emma Francis, and Murray Pratt (Hampshire, UK: Ashgate, 2002), 77, original emphasis.

24. Bartky, 'Foucault, Femininity', 149.

25. Sally R. Munt, 'Orifices in Space: Making the Real Possible', in *Butch/Femme: Inside Lesbian Gender*, ed. Sally R. Munt and Cherry Smyth (London: Continuum), 201.

26. Munt, 'Orifices in Space', 205.

27. Sarah Moore and Simon Breeze, 'Spaces of Male Fear: The Sexual Politics of Being Watched', *British Journal of Criminology*. Advance access published August 9, 2012, doi:10.1093/bjc/azs033, 6.

28. Bartky, 'Foucault, Femininity', 148.

29. Bordo, 'The Body', 99.

30. Munt, 'Orifices in Space', 203.

31. Chris Shilling, *The Body and Social Theory*, 1st edn. (London: Sage Publications, 1993), 7–8.

32. Ahmed, *Queer Phenomenology*, 23.

TINY BLAST

Peter Gizzi

Just a small song with a dash of spite.
A tiny thistle below the belt.

That's it, you know,
the twinge inside this fabulous cerulean.

Don't back away. Turtle into it
with your little force.

The steady one wins this enterprise.
This bingo shouter. This bridge of sighs.

And now that you're here be brave.
Be everyway alive.

6

HOMINES APERTI **AND MATTERS OF CARE**

It is impossible, however, to deal adequately with the problem of people's social bonds, especially their emotional ones, if only relatively impersonal interdependencies are taken into account. In the realm of sociological theory a fuller picture can be gained only by including personal interdependencies, and above all emotional bonds between people, as agents which knit society together.

—Norbert Elias

This chapter explores different modes of caring for bodies in public toilets that explicitly challenge one or more rules of the *homo clausus* triadic intra-action order. The intra-action order seeks to stabilise *homo clausus* body-identity by keeping the imagined borders of the body as controlled as possible. While caring for others in many aspects of public life is socially laudable or acceptable, caring practices within the confines of public toilets are often transgressive of social norms as organised by the triadic intra-action order (TIO) and highly problematic. Crucially, the expressions of care I highlight in this chapter expose potential fissures between the binds of (*homo clausus*) individuality and human sensory-embodied desires and needs for care, closeness, and intimacy with other people (*homines aperti*). In this chapter I argue that this is because 'caring in toileting' overtly exposes the inherent openness and interconnectedness of bodies, highlights their vulnerability, and reveals how the monadic confines of *homo clausus* norms are contingent and frail rather than universal and stable. Toileting practices necessitate that bodies are open, yet openness is antithetical to a stable *homo clausus* identity and thus any acts that directly socially reveal and sustain this openness are often seen as despicable. Care does happen in public toilet spaces, however, and I want to suggest that this exposes opportunities for valuing the body as dynamically living and as evidence that it is possible to understand individual identity instead via *homines aperti*—that is, showing how identity is formed through direct connection with, rather than separated from, other bodies. This is akin to Frank's model of communicative bodies, at least insofar as its descriptive nature and proposition of an ideal bodily ethics.[1]

However socially challenging it may be for those individuals who live through modes of care, it is important to recognise that what I shall analyse as this move from *homo clausus* to *homini aperti*, as experienced through material-discursive practices and sensorial engagements, is itself an opening for new ways of sensory-embodied becoming. By giving non-judgemental attention to practices which are socially understood as non-normative, but that actually occur regularly in mundane, daily practices, we can bridge our conceptualisation of the 'human' with those experiences of bodily living that are ignored or systematically neglected in that conception. This is where onto-epistemology can be usefully highlighted for better integration into our ways of experiencing, knowing, and understanding. For example, instead of applying and then analysing a socially prescribed, pre-digested sensation or reaction to an experience (e.g., why some women feel threatened by the presence of a butch lesbian), I use the data presented in this chapter to expose where we can break habitual conditioned judgement by better understanding how-where-when identity is experienced as interdependent and begin to move towards new practices of sensory-embodiment.

The empirical data within this chapter focuses on mundane social practices of daily caring which are expressed in multiple forms, reoccur, and fall into roughly three categories. These are: protective care, collective care, and bodily care. This chapter includes data from men, women, queer, and trans individuals, with a range of sexualities, as well as queer and heterosexual couples, including some who are parents of young children. The data offer multiple perspectives concerning how, why, and when care happens in public toilets. The examples explored here are not concerned with how people maintain the *homo clausus* intra-action order in an effort to maintain bodily boundaries (e.g., how a mother may make a bed of toilet paper for her young daughter to sit on thus trying to control what her body can come into contact with, an example of rule three of the TIO, manage your boundaries), but rather how bodies are cared for beyond or in direct contention with those practices allowed by the intra-action order. Therefore, through a confrontation with and acceptance of themes related to vulnerability— often by ignoring or moving beyond socially instituted feelings of bodily fear, anxiety, shame, and embarrassment—this chapter will show how the borders of *homo clausus* identity can easily break apart and/or be actively dissolved in daily life, allowing for new forms of embodiment and social cohesion. This is a move towards a new ethics of being bodily.

Practices of care in mundane, daily circumstances embrace the leaky, unstable, abject body through honest acceptance, expression, and acknowledgement of bodily needs. The material-discursive practices elucidated in this chapter begin to point to the possibilities and potentials which are available to everyday embodiment but which are often precluded through social patterns of use, like those mandated by the triadic intra-action order of public toilets (TIO). While caring for bodies in public toilet spaces may at first seem odd, disgusting, or taboo, when we situate these practices within the paradigm of embodied knowledges we can being to see how such knowledges are devalued in society and reflected in ourselves. By understanding the circumstances where caring practices are destabilising to *homo clausus* identity it becomes possible to build an awareness of where sensory-embodied becoming can be released from the rigid boundaries of the monadic self-body. Put simply, practices that can be characterised as typical of *homines aperti* can help us locate thresholds for becoming-other.

PROTECTIVE CARE

Since public toilet spaces are intensely straightened and maintained through the TIO, there are many opportunities for and threats of transgressive action. Often someone does not desire to transgress the TIO overtly, but is read by users of the space as inherently suspect and threatening simply because of the way that person embodies and performs sex-gender-sexuality in this 'private' space. For example, while some queer and/or trans people may not want to disrupt or disturb anyone else who is using a public toilet, it is often difficult for them to avoid doing so because the spaces are so rigidly maintained according to heteronormative *homo clausus* ways of being that are reliant on the visual for affirmation. Queer and trans people often simultaneously feel *both* threatened and threatening in public toilet spaces, and because of this paradox of threat, they often put great care and attention into using public toilets. In the examples below, the practices of care are both protective and preventative, expressed in efforts to try to avoid a potentially dangerous or personally damaging experience.

Beck is a fifty-year-old butch lesbian whose body-identity is constantly policed when she uses public toilets. While her friends have encouraged her to use men's public toilets (an even more risky option) instead of women's, she maintains that she does not want to. This is an example of protective self-care as she is protecting her sense of self both by choosing to do what she finds most comfortable and by avoiding the very real potential threat she could face if she used men's toilets and was discovered to be a queer women. She explains:

> Everyone says 'just go into the men's room; men won't say anything' and I'm like fuck no. I'm just not. I don't want to go in the men's room! They're dirtier for one, for the most part and I feel like I'm not supposed to be in the men's room so I don't want to do that.

While Beck may look more conventionally masculine than feminine in her adult presentation of self, she is a woman who is subject to the same deeply instilled feelings and fears about women's access to men's public toilets that heteronormative women generally are. She feels strongly that she does not belong in those spaces and that they are dirtier and smellier than women's public toilets—a common assumption (and misconception) among both women and men. Her desire to only use women's public toilets, despite her butch appearance, points to how deeply ideas and beliefs about sex-gender are embodied from an early age and the ongoing historicity of the body that informs ways of living. While the current (adult) expression of her identity may not agree with the early socialising into femininity she experienced as a child, those messages about her body only being with other women's bodies while she engages in the intimate act of excretion remains. Beck may more easily and readily identify with men and masculinity in her bodily gender-sexuality expression, but when it comes to her sexed body, she still feels she needs to be with other women for this act, despite the constant strife it causes her. This highlights the importance for Beck of maintaining her ability to be able to use women's public toilets despite the dubious status of her feminine legibility.

One of the most successful methods she has developed for using a women's public toilet without harassment, or at least with much diminished social attention, is through the care of her girlfriend. As Beck explains:

> Sometimes I don't want to be bothered [by other women in public toilets] and my girlfriend will say 'do you want me to come in with you?' and I'm like 'Yes! Please!' And she'll go in first and then I don't have to say anything because she's got a bigger evil eye than I do! She's very femme and it makes me feel like I don't have to say anything because she'll just take over and then I don't have to confront anyone looking at me while I'm trying to go to the bathroom. Or she'll turn around and talk to me as if, ya know, it is no big deal [that this very masculine-looking person is in here]. So maybe that is

it, when I'm with her, it makes me feel like I don't have to confront it at all. I don't have to deal with it alone or at all. She'll protect me.

This practice of protective care is directly tied to Beck's experience of the legitimacy of her body-identity. The clearly feminine status of her girlfriend normalises, or at least legitimises, Beck's presence in the space. While this practice of care may be seen as taboo because it exposes an openness and connection between bodies that jars with the heteronormative *homo clausus* social codes of the space—i.e., a clearly feminine woman openly accompanying a masculine-looking body into a sacred feminine space—it is a vital example of people as *homines aperti*. As Elias explains, 'people look to others for the fulfilment of a whole gamut of emotional needs', and the physical expression of the emotional bond between Beck and her girlfriend speaks directly to this point.[2] While this may not be a major event, it is a small expression of protective care which is predicated upon an emotional-physical bond. It is important to note that while it is a generally accepted fact that women often use public toilets in friendship groups, this example is quite different from that. This form of shielding is protective *from* women and especially those in groups who can be even more threatening to someone like Beck. Women who use public toilets in groups as a social practice is a form of *homo clausus* normativity and something that most queer women I interviewed claim to not understand and never engage in. It may be a caring practice, but it is a wholly normative one and thus not of interest here. When women use the toilet together in groups it both socially eschews the need to actually engage in excretion, because it is couched as a social activity to others, and detracts from any sensory-perceptual evidence of the actual excretory process—since more sounds from more bodies means less direct responsibility for sounds and smells individually produced. It is a form of social policing in the form of sociality. Women using public toilets in groups is a public and private heteronormativising, which may have some benefits of protection and solidarity, but it is not disruptive to *homo clausus* norms.

Alternatively, Beck and her girlfriend's practice of care shows the utter openness of sex-gender-sexuality in its simple yet powerful ability for an expression of femininity to be overtly protective over and in defence of masculinity. Beck's 'femme' girlfriend, through the strength of her socially legible femininity, is able to both protect Beck's embodiment of masculinity and support her unconventional sensory-embodied identity. This works at least in part by other users locating a reflection of socially accepted representation of heteronormative femininity in Beck's girlfriend, which grants her access to the space and an implied understanding that she knows the rules and codes of it. Thus she is able to use her normatively read body in an act which is non-normative. If Beck's girlfriend wasn't 'femme' I suspect this shielding practice would be much less successful, creating even more problematic confrontations.

While protective caring practices are much less common in men's public toilets they do occur. Joseph, a twenty-six-year-old queer man, shared this story with me about himself and his thirty-one-year-old transgender friend Jason (who also partook in this study) about engaging in a shielding practice, like Beck and her girlfriend. Jason is more butch and hetero-masculine in appearance than Joseph, who is on the slighter side and often read by people as 'gay'. Joseph explains:

> The bathroom was in this club, which was filled with these really intense hetero guys and we just swapped door duty—because the bathroom wasn't explicitly for one person, but it was pretty small. . . . I mean I did it for protection, and felt like it was important to protect Jason too. And to combat being aggressively accosted or having my space invaded by some hypermasculine Long Island dude.

Here Joseph and Jason protect one another by taking turns using the toilet and acting as 'door security'—standing at the door and not allowing any other men to enter the space. A similar story was shared by Justin, who is a thirty-three-year-old transguy. He says:

> One time when I was playing a show [with my band] I was with a trans friend and we blocked the men's bathroom off for each other, we guarded each other. I really, really liked this. It was like, really empowering. Yeah, I totally liked it. It was this acceptable protectiveness of my dude. I guess maybe it's a territorial thing, it just felt empowering to keep people from entering and knowing he [trans friend] got his privacy too.

These practices of protective care over the ability to safely carry out intimate bodily processes are expressions of masculinity rarely captured publicly. While these sorts of practices are considered non-normative by the TIO and heteronormative *homo clausus* masculinity (surely they would be labelled as 'gay' according to the TIO) they are not experienced as emasculating by those men who engage in them. On the contrary, they are described as empowering and important masculine acts. Like in Beck's story, the *act* itself supports a heteronormative construction of sex-gender-sexuality, but the intention and experience of the act is utterly disruptive to *homo clausus* norms. The act of protection through a sort of territorialisation can be understood as quite normative, but the desire to protect another man's excretory needs by creating a safe environment for them is decidedly non-normative.

Similarly, in subsequent interviews and social conversations with straight men, when I would proffer this practice of care as an example of my findings, many seemed fascinated and even envious of this expression of masculine, protective care for friends. In these conversations, it was as though labelling something as queer or gay from the beginning nullified the threatening power of being identified and labelled, called, or thought of as gay by other men. While generally men, in order to maintain the TIO, are concerned not so much with appearing *straight* as with making sure they do not do anything that may make someone think they are *gay* and thus suspect and threatening, when it came to hearing these stories of protective care by queer and trans men, the general sentiment was that these were stories of liberation (!) from heteronormative masculinity, not examples of shame or fear. So while the heteronormative men I spoke with, upon hearing these stories, said it was something that would 'never happen' between straight men, they also did not find this behaviour threatening or repulsive in any way. This is unsurprising, because while this sort of behaviour completely violates the rules of the TIO, it simultaneously invalidates the entire premise of needing to appear straight in the first place. When men simply are not so concerned with not appearing gay, the power of the TIO is radically diminished, since the entire premise of *homo clausus* normativity is based upon the closed, rational boundaries of heteronormtive masculinity. That is to say, men, by simply engaging in different practices—perhaps spurred by hearing of those practices of queer men—may feel less compelled to so intensely control their own bodies according to the TIO. This is how fear, anxiety, shame, and embarrassment in conjunction with the intra-action order begin to lose their persistent power; by taking small risks in breaking their embodied habits, men can begin to be differently embodied. This protective care works because, while hetero men may not find this queer protective care overtly abject, the way that the TIO works is less by men being concerned with what other men are doing and more concerned with their own behaviours.

Joseph's, Jason's, and Justin's protective care points to the inherent openness between bodies, and particularly between men's bodies emotionally and physically. The affective bonds expressed by these men point to 'the possibility of there being very strong emotional bonds of many kinds without any sexual overtones' as basic and integral to human relationships regardless of sex-gender-sexuality.[3] Furthermore, these expressions of care, as understood by those men rigidly tied to the heteronormative *homo clausus* TIO exposes a potential fissure between the binds of masculinity (*homo clausus*) and human sensory-embodied desires and needs for care, closeness, and intimacy with other people (*homines aperti*). This is an example of a process of individuality as explained by Elias. He says, 'Biologically determined instincts are still present, but they can be greatly modified by learning, experience, and the processes of sublimation.'[4] That is to say, these practices of protective care expose a bodily desire to be close and connected to people, through overt actions of connectedness that many men have learned to sublimate or express in only very rigid, socially deemed appropriate ways. Protection of this sort speaks to an ethics of materiality that is beyond mere biological control and management. Rather it speaks to the active nature of materiality that informs a desire to be close and connected to other bodies even if that requires a transgression of heteronormative identity. Protection is a practice of care that overcomes the social feelings that help maintain the sensation of atomisation and helps highlight the intra-active nature of bodies. These are acts of solidarity and hospitality, with oneself and others, which value differential being not over and above representational sameness. This is similar to practices of collective care which are even more intimately related to the bodily desire for human connection as it bumps up against individual anxieties surrounding the maintenance of *homo clausus* boundaries.

COLLECTIVE CARE

Practices of collective care expose the social and interpersonal struggle some people face when opening the body in excretion cannot be a purely individual act. The struggle between maintaining individual adult status necessitated by the TIO, while requiring or highlighting the need for assistance in public toileting, brings bodily fear, anxiety, shame, and embarrassment (FASE) to the fore. When practices of collective care—those entanglements that implicate personal assistance in the acts of excretion—intervene in bodily FASE there is an opportunity for those emotions to be released and a new opening for fostering bodily connection and compassion can be revealed.

Frankie, who is twenty-five and queer, shared one story of collective care with me about her and her girlfriend. While her girlfriend generally has bathroom anxiety and does not feel comfortable receiving care in public toilets, one incident quickly changed their usual dynamic in these spaces. Frankie explains:

> We [me and my girlfriend] had never been in the bathroom together
> while pooping, and one day, she [my girlfriend] was in there pooping
> and I heard a huge bang and she had fucking fainted while going to
> the bathroom, or right after and I had just heard the bang and it
> scared me! So I went to the door and she didn't respond, so I just
> went in and she was naked, kind of half naked passed out on the floor
> and had hit her head, and it was really scary! And all of a sudden that
> awkwardness [of being in the toilet together] disappeared. I don't
> think I'd go in with [her] today to poop, but that day and the next
> couple days, I was really aware and okay with being in there with her,
> and it is funny how the second I thought [she] could get hurt again,
> [she] didn't care that I was in there . . . that boundary broke down
> when there was a seeming necessity for me to be in there and now
> we're way more comfortable with our bodies bathroom stuff general-
> ly.

In this example, a couple who had never shared this particular aspect of
their body-selves quickly and easily overcame any FASE they had previously held in association with toileting and allowed a new bodily openness and connection to happen between them through an act of care
and compassion. That incident has had an enduring effect on their
relationship and on their embodiment, enabling them to now be 'more
comfortable' with their bodies generally and particularly with toileting
together. This highlights that the rules of the TIO are socially contingent and can be overcome and is in contrast to Butler's (1993) and
Foucault's (1980) understandings of discursive power which are unable
to account for how sensory-embodied individuals can challenge or
change the workings of power (i.e., how materiality is related to discourse). The above story highlights how agency happens intra-actively
through material-discursive practices, that is, how new ways of being
are entangled with new ways of understanding. Put simply, when personal habits and social propriety are ignored or overcome, a threshold
for differential becoming can emerge.

The preceding story is in stark contrast to an earlier experience in
Frankie and Natalie's relationship. Natalie, Frankie's girlfriend, who is
twenty-four and queer, told me another story when she was unwell in a
public toilet. She recalls:

> I'm remembering at the Bellhouse [a large performance-dance-mu-
> sic venue] I had a stomach flu and I was in the bar's bathrooms for
> like an hour—we were there for a show—it was a two-stall bathroom
> and I just could not physically get out of the stall, I was unable to
> move, cause it was coming out of both ends and [Frankie] kept
> coming in [to check on me]—there was a huge line of people and I
> could hear women saying 'someone's been in there for so long!' I felt

really bad, [Frankie] kept coming in and asking if I needed help, and
I just felt so awkward and embarrassed, I just kept saying, no it's
okay, it's a public restroom. . . . I just felt really bad receiving care in
the public bathroom.

In this story, Natalie expresses her rationalised social discomfort with
receiving care in public. Her story highlights why it is so horrible to be
ill in public toilets under the *homo clausus* regime of the TIO; you can't
relax, you feel on display, you feel childish, you feel completely out of
place, all of which negatively intensifies everything you're experiencing
to begin with. Natalie's FASE about publicly admitting the need for
help, for requiring care, meant that she struggled, ill, embarrassed, and
alone, for the sake of maintaining whatever aspects of the *homo clausus*
TIO that she could. I'm not suggesting that maintenance is necessarily a
conscious process but rather an unfortunate, unnecessary social expectation that denies one the right to ask for help, to take up space, and to
be in a state which is not up to the *homo clausus* ideal because we
generally do not understand or realise that these emotions are not natural but rather socially instituted or we do not have the tools to overcome
their oppression.

While Natalie thought that actually receiving care would be even
more embarrassing than denying it, using a huge amount of time and
space, and being sick alone, as evidenced in the first story told by
Frankie, this was not the case. Instead of allowing her rational FASE to
control her bodily needs and desires, when she received care from
Frankie after passing out she realised that it was much less problematic
than she thought it would be. For Natalie, it was only through overtly
experiencing her body-self in the *receiving* of care that she is able to
overcome her rational FASE and feel more comfortable in her body
and with Frankie's body too. This, again, highlights how material-discourse is intra-acted and inherently material—that is, entirely entangled with sensory-embodiment.

The next two stories explore collective care from a slightly different
angle. While the stories are told by those who are socially and emotionally close to those giving and receiving the care, they usefully highlight
the *homo clausus* anxieties felt by those who merely witness the acts of
care. The first story was shared with me by Monica, who is a forty-six-
year-old lesbian. She explains:

It [our conversation about sex-gender-sexuality policing] makes me
think of my step-father who had a stroke ten years ago. So he is an
over-sixty-five-year-old man who can't go into the men's room by
himself [because of limited mobility resulting from the stroke], so my
mom has to take him into the women's room with her and people will

say shit to him! They still think, they still think he is an intruder or a pervert or something! My mom is there, helping him walk, he's got a cane, she is clearly helping him, he moves very slowly and people will still say things to her about him being in the bathroom! And it is just like, oh my god, close the door or wait until he leaves to pee if you're that embarrassed about it! It is so frustrating that people are so freaked out about this!

The deep frustration expressed in Monica's story is palpable. It seems difficult for her to understand how people can be so ungracious when faced with this display of collective care. While she explains this as the embarrassment of those women who hound her mother and stepfather, there is much more at stake here than simply women being embarrassed about a man hearing them pee.

Her story brings issues and ideas concerning sex-gender-sexuality, age, (dis)ability, *homo clausus* bodily boundaries, masculinity and practices of care to the fore. Despite her step-father's clearly legitimate need for care by his wife, displayed in his slow comportment, use of a walking aid, and presence in the 'wrong' public toilet accompanied by a female aide, women still feel threatened enough by his mere presence to harass them. This daily reality for Monica's parents reveals a troubling feature of *homo clausus* selective attention.[5] It seems some users of women's public toilets are so deeply offended by the presence of men in 'their' space, they fail to recognise *and value*—put another way, to *empathise* with—the fact that his needs are clearly different from those of an able-bodied person and that his receiving of care is clearly legitimate. We can speculate that witnessing this act of collective care is troubling for some women because, despite the maternal trope of femininity, it is generally understood that when it comes to issues of physical weakness it is men who are expected to provide support for women. Put simply, the display of an adult feminine body providing care in the form of physical strength to an adult masculine body within a public toilet fundamentally challenges many *homo clausus* assumptions of sexed-gendered bodies, which tend to be shored up and *stabilised in the toilet space*. So while women may realise that a disabled man may require care by noticing his comportment and use of a walking aide, they do not find his need legitimate enough to be accepted without harassment. This act of collective care draws out socially reproduced FASE and points to where *homo clausus* constructions of identity fail to be open and amenable to the embodied possibilities of daily life. As women are socialised to feel afraid of men in women's public toilet spaces, some are unable to look past the assumed presence of a penis (which must mean danger) in the space and compassionately connect with the practice of interpersonal care happening before them. Such FASE is a clear display of where assumptions of sex-gender are delimited onto the body via *homo clausus* identity and allowed to precede the ongoing material reality of sensory-embodiment. It seems some women so fundamentally believe that men's experiencing of their bodies and their approach to the bodies of 'others' is (and should remain) so different from their own that they are unable to connect to a very basic human need for care. That is to say, some women may have so much invested in the construction and reproduction of heteronormative femininity that they are not able to understand an equally large investment two people may have in one another. Similarly, as Elias explains, 'People's attachment to such large social units is often as intense as their attachment to a person they love.'[6] It seems in Monica's story both cases are present; some women are so connected to their social unit of hetero feminine gender that they are unable to reconcile, with their own sensory-embodiment, an expression of love between two people because of the location in which it takes place—a location that is arguably instrumental in the reproduction of that social unit.

The last story of collective care is similar to Monica's in many ways, but happens from the reverse. It reveals, instead of women's fear of men's aggression and danger, men's concern for aggression or abuse towards young girls. As elucidated in the previous chapter on the triadic intra-action order, *homo clausus* material-discursive practices construct bodily boundaries in equal yet opposite ways, resulting in binary sex-genders that are reliant upon one another for stability. Therefore, where Monica's story exposes women's fear of being victimised by men, this next story exposes men's fear of victimising young girls. Deborah, who is a forty-one-year-old, heterosexual woman, shared this story with me about her family:

> So we [my family] often use public toilets, but it is also an issue for us because obviously it's my husband [who is with them]—we have [three] girls [aged two, six, and seven]—he is most often out with them [because I work full-time] and he finds it rather difficult because he doesn't really want them to go and use the toilet on their own in the ladies [women's public toilet] but it isn't always particularly suitable to bring them into the gents [men's public toilet] but that is usually what he has to do. . . . We have a particular issue with where they go to do gymnastics; all the girls—there are obviously toilets there and adults use the gym too—and, well, he used to bring them into the men's changing room to use the toilet and so they can get changed but he was actually asked not to. They said, 'could you not bring them into the men's room because they might see other men using the toilet'. Now, this situation is very difficult! I mean,

what is he supposed to do? So it is quite an issue that I hadn't really thought about before but that is difficult for dads. It's hard to deal with, you know. It is quite different for women; you can take a little boy into the ladies toilet and you've got a cubicle there, so it's not really an issue, I think. So it must be the privacy thing, because my husband wouldn't let the girls see anything he wasn't happy for them to see, but the people at the gym didn't want them to go in there. I don't know, I'd like to question them [the people at the gym] about that, because I'd like to know if they thought other men might be encouraged to do it [bring their daughters into the men's toilet] or are they more concerned for the children? But he was there, he was there! He would have been around, he would look after them. He wouldn't have let anything happen to them. Yeah, it made me, it really surprised me that they said that [he couldn't bring his daughters into the men's room anymore]. Well, I suppose it wasn't really our problem; he [my husband] is happy to take them into the gents and he will look out for them, but I guess it makes other people uncomfortable.

Deborah's story is quite illuminating because it touches on many different sex-gender issues at once and is something rather simple, yet surely plagues many families.

People are uncomfortable by the mere idea of young feminine bodies in a masculine space and the potential of young feminine eyes *seeing* men's bodies; young masculine bodies present in women's spaces do not, in any way, represent the same socio-cultural concern or stigma as the oppositely gendered configuration. At first thought this may seem ironic, considering Monica's story, but upon closer consideration it shows how basic ideas of binary sex-gender are deeply instilled into even the youngest bodies in social life. Here young feminine bodies are expected to be victimised by masculine bodies by their mere presence in the space. The concern assumes that seeing a man urinating or just seeing a penis is inherently violent to young female sensibilities (that is, to see a, most likely, flaccid penis is somehow violent, unless it is belongs to their father or brother, in which case that is generally accepted), and thus they are treated *as victims* of male violence in the name of avoiding such possibilities of harm. In my interviews I heard many stories from women who accidentally used men's toilets as children and have lasting positive and even affectionate memories of seeing men urinate; as though the vulnerability of men in that position is to some degree comforting. Correspondingly, this dynamic reproduces the assumption/expectation that men are violent to women; it is part of a larger social process that teaches girls to be afraid of men and that they are weaker and inferior to masculine bodies. As Twigg drawing on Connell (2000) states, 'Hegemonic masculinity constructs men as sexually predatory; and limits are placed on male access to bodies . . . ' whereas 'women are allowed greater leeway in performing the transgressive acts of bodycare without their being constructed as threatening or sexual.'[7] Young masculine bodies, while not thought to be able to victimise adult women's bodies, are not treated as victims in women's public toilet spaces and their presence is not socially problematised. My goal here is not to reproduce these cultural tropes but instead to point out how *homo clausus* material-discursive practices impact the materiality of bodies in their reproduction of heteronormative sex-gender.

In this example specifically, *homo clausus* individuality is invoked in order to deter a practice of collective care between an adult man and his three young daughters, an example that re-inscribes feminine bodies as victims and masculine bodies as perpetrators of violence. The managers of the gymnasium leave Deborah's husband with few options regarding the mundane care of his daughters. Since he has been asked to not bring his daughters into the men's public toilet and he cannot access the women's public toilet, he now sends them into the women's room unsupervised and stands by the door, waiting for them. The irony of this situation is that the FASE surrounding bodies in public toileting scenarios takes precedence over the caring of a father for his daughters. It removes his agency as a father and as a man engaged in a caring practice. Deborah's daughters now have to use the women's public toilet on their own—certainly making them more vulnerable than if they were with their father—and her husband now has to loiter around the door to the women's public toilet to try to make sure his daughters are okay—certainly rendering him suspicious and potentially threatening in the eyes of other women. The social stigma surrounding young female bodies with male bodies transforms a masculine caring practice into a practice that re-inscribes men as suspect, untrustworthy, and threatening to women (with an underlying current of sexual violence). This is a clear example of the entangled nature of intra-acted material-discourse in action.

The whole situation is compounded by the fact that the girls' father is their primary caretaker while their mother works full-time. While it is socially acceptable and even laudable for fathers to act as primary caretakers in young families—a role that challenges heteronormative gender stereotypes—when it comes to the practice of fathering in public, fathers are still plagued by *homo clausus* gender stereotypes based on the sex of their children which may imply that men are not suited to full-time care. (There is an underlying implication in this story that the father is unable to keep his children safe around other men.) The gym example exposes how in public toilet spaces it remains difficult for men to act as the primary caretaker of young children who are gendered female. This is another instance where *homo clausus* constructions of individuality cannot account for realities of the open, interconnectedness of people in their daily activities and how people do not necessarily follow conventional sexed-gendered constructions of heteronormativity in living sensory-embodiment. As *homo clausus* bodies are threatened by the overt openness of bodies, it is vital to also explore the intimate bodily care people engage in independently during acts of excretion, which clearly have social ramifications. As evidenced in the final section of this chapter, when bodies are open they are no longer individual (and bodies are always already open).

BODILY CARE

In order to make the case for bodies as having boundaries that are not stable or static, but rather involved in an opened-ended becoming (*corpus infinitum*), we must first give attention to the material openness of bodies. While using a public toilet for the literal act of excretion may be thought of as an entirely independent act, the materiality of the body cannot be contained to one's flesh as it produces sounds, smells, and wastes which vitally transgress imagined bodily boundaries. If this were not the case, the power of bodily FASE to control material-discursive practices would be radically diminished. Thus, even when using a public toilet alone, as a singular person, bodies are always already intra-acting because of their inherent co-presence. The following examples aim to draw attention to this point. They show how when people are alone their bodies are still entangled with other bodies in the space and how this care is an undervalued and necessary form of 'dirty' bodywork. According to Twigg, 'The term "bodywork" has commonly been applied to the work that individuals undertake on their own bodies, often as part of regimens of health and wellbeing.'[8] I draw on Twigg's insights on 'dirtywork'—i.e., dirty bodywork—throughout this section to elucidate how such basic and universal caring practices as excretion can be understood as utterly despicable and at odds with one's 'self-image'. The stories shared below are bound up with people's working lives and expose how when one tries to take care of their needs in a space where bodily intimacy and openness is at best not valued and at worst problematic they are ostracised.

Ford is a twenty-four-year-old queer (ftm) trans man who, at the time of our interview, had undergone top surgery and been on testosterone for many years. His self-body identity is read by others as male and he is rarely questioned about his sex-gender status. If anything, people assume he is a gay man (which can often be problematic for him as he feels he is still a 'gender minority', solidly identifies himself as 'trans' and often works in genderqueer and trans advocacy). While he is 'passing 100 percent of the time' he still has deep concerns about using public toilets. His experience of bodily FASE extends beyond the confines of a public toilet, and for ease of explanation his story requires some context. He explains:

> Bathrooms were a huge issue in my transition and they still invade my thoughts. Before, during, and after my transition I have had nightmares about bathrooms. . . . In them [the nightmares] the stalls have no doors, or the door on the stall continues to shrink exposing me more and more every second, or I walk into the women's room and everyone starts screaming at me.

Before and at the very beginning of his transition (to masculinity) he would go out of his way to use a unisex bathroom, to avoid using women's public toilets where he felt he didn't belong. At that time he was masculine enough to be policed in women's spaces but still too feminine-looking to comfortably and confidently use men's public toilets.

In order for one to begin taking testosterone, a lifelong treatment for trans men, one has to first be in therapy for, at least, several months, and be 'presenting' (one's body-self) as a man daily. While Ford passes as male now without any issues, this was not always the case. He shared this story with me from the beginning of his transition:

> I started T [testosterone] close to the same time I started a new office job. One day I was in a stall in the men's room and there was one other person in the bathroom, and on this person's way out they stopped at the door and shouted [at me] 'This bathroom is for men only!' After this incident and speaking to HR [human resources] I found out that there had been a lot of complaints made about me using the men's room but the HR person never said anything to me and basically they didn't know how to handle it. . . . Later I found out that there was a trans woman who worked in the same building a year before who also had a lot of issues using the women's room and she ended up leaving her job because of it. This situation was hard because I needed a bathroom to use and I didn't feel that I 'belonged' in the women's room, yet other people felt I didn't belong in the men's room. To make matters worse, at this job I often had to stay late and work and my boss told me not to use the bathroom after business hours. So, my boss was making me stay late to work but told me I couldn't use the bathroom!

Here, despite Ford's status as a trans man who was taking male hormones, presenting as male every day, and engaged in the material-discursive practices of masculinity in public toilets and elsewhere, his body in excretion was still highly problematic for some of his co-workers. This instance shows how even the extremely mundane and universal act of bodily excretion can become an intense and intensely charged caring practice for someone. That is to say, while someone who is not regularly hassled and denied access to space may never think of toileting as a fundamental right to self-care, the situation is wholly different for those who are differently abled and/or face prejudice in public life. While the men who shared the public toilet with Ford did not necessarily feel physically threatened for their safety by his presence, his presence caused enough FASE in some users of that space to prompt complaints about him; and they were not taking into consideration this basic need and right to care for his body. [9] As the *homo clausus* TIO in men's spaces operates by keeping men mostly concerned with their own bodily actions and not those of others, Ford's story suggests that the psychical attention some men could not resist giving his body meant they were transgressing the TIO, which inherently has negative implications for the status of their heteronormative masculinity.

Furthermore, this experience was surprising to Ford because in the other spaces of the office there was no indication that anyone was overtly uncomfortable with his being there. It was only in this space of literal openness, when people gave their bodies directed and careful attention, that his body became problematic enough to warrant complaint. This shows that even those who may imagine their bodies are closed and sealed can feel open and exposed—made vulnerable by bodies that are not conventionally masculine. This vulnerability may be compounded by the act of bodily care—that is, an act of self-love and labour—as masculine bodies are not typically inscribed (or easily accepted) into body-caring practices and roles. [10] As Twigg explains:

> There is a complex set of reinforcing influences that together construct bodywork as female. First, these are tasks that are naturalised in the bodies and persons of women. Women have traditionally represented the Body in culture (Jordanova 1989; Lupton 1994). They have been presented as more bodily than men, bound up in and defined by the processes of reproduction, and prey to the shifting tides of emotion. Women also represent the Body in terms of male desire, the form of desire hegemonic in culture. They thus come to represent sexuality itself, something that can be controlled through the control of women's bodies. Confining desire (at least in its legitimate forms) and the needs of the body to the domestic sphere allows

the public world to be constructed as disembodied, rational and male.[11]

Toileting as a form of bodywork for men is only acceptable through the intensely individualistic and rational nature instituted by the *homo clausus* TIO—material-discursive practices which have direct implications on male desire—as men are trying to avoid being read as anything but heterosexual while in public toilets. When there are bodies present that blur the sex-gender binaries and highlight the bodily caring aspects of excretion, it may draw men's attention away from their task at hand to the other bodies in the space, which may result in those heteronormative men feeling uncomfortable and threatened by *their own* transgression of the TIO. The men in Ford's story are not directly threatened by Ford—Ford *makes no threats*—but are instead made to feel endangered by their own cognitive schemata, their own thoughts about Ford's body.

Seemingly, it can be difficult for some people to imagine different, non-normative bodies in basic modes of care. It is interesting that the person who shouted at Ford waited to do it on their way out, when Ford was indisposed, locked in a toilet stall; a rather cowardly expression characteristic of *homo clausus* FASE. Furthermore, the human resources person's total inability to engage the situation as something that needed attention and warranted a conversation with Ford is upsetting, as is his boss's insistence on his working late but not being allowed to use the toilet. With situations like these it is unsurprising that queer and trans people are victimised in toilet spaces when they're simply trying to care for basic bodily needs. Luckily, this attack on Ford was verbal, pusillanimous, and not physically violent; though it is troubling that he had to be victimised in this way in order to get his HR person to get involved in looking out for the welfare of employees. While it does not excuse the neglect by Ford's HR person, it is not wholly surprising. Toileting practices are a form of bodywork that is 'closely connected with the negativities of the body, and these are aspects that modern culture tends to shy away from, in analysis as much as in day-to-day life.'[12]

In this example Ford was simply trying to take care of his bodily needs; he was not doing anything devious or overtly transgressive, he was simply using a public toilet for excretion. While those basic needs are generally accepted as the same for all bodies that use public toilets, his excreting body (while in the presence of other open, excreting bodies) was somehow outside of this general acceptance and too problematic for some people to handle. Even though his excreting body was not something his co-workers were ever confronted with directly (visually or physically as he always used a stall with the door shut and locked), the rational construction of *homo clausus* identity, as a stable sexed-gendered body, cannot be reconciled with the reality of Ford's presence in the toilet space. Ford's unconventional sex-gendered body inherently challenged some men's sense of masculinity, stirring their own bodily FASE, simply because his body does not agree with how masculine *homo clausus* bodies are expected to be—that is, the idea of what a man's body is and should be. In this instance it seems there is fear and anxiety linked to what men may see if men were to 'accidentally' look at Ford's excreting body (certainly breaking all rules of the TIO). The thought being, 'I know what I would see if I saw a trans man's body and I know that would disturb me'. While, as far as we know, no men spied on Ford whilst he was in a toilet stall, the feelings and actions of those men who complained about his use of the space are seemingly directed by the rational *homo clausus* mind, not the living, perceiving capacities of sensory-embodiment. Whether they like to admit to it or not, in public toilets men's bodies are interrelated and reliant upon one another for the maintenance of not just the TIO, but masculinity generally. The reasons for the neglecting of interconnectedness is typical of *homo clausus*—as they are 'deeply rooted in Western culture and in the evaluation of the bodily, particularly those aspects of the body that run counter to modern preoccupations with autonomy and individualism.'[13] Even though this was not a positive experience of connectivity for Ford, it is nevertheless apparent in his story that bodies are open and interconnected, that they have the power to affect the feelings and actions of others. Situations like these continue to haunt Ford and can seriously deteriorate a person's life and, as shown in the next stories, this is true not only for trans individuals.

Miriam is a twenty-five-year-old queer woman who shared two work-related public toileting stories with me. The toileting space at her place of work was atypical insofar as it was genderqueer.[14] The space did not have a stable gender but rather swiftly conformed to whoever was using it according to a sign on the door. The public toilet space plays a vital role in the elucidation of her experiences and thus it is important to have a clear sense of what it was, how it was used, and how people generally felt about it. Miriam described it to me in the following way:

> I had an interesting bathroom situation at work where we had a gender neutral bathroom. It was a single bathroom [one room, with one door] but inside it had two stalls with shared sinks and it was meant to operate as a gendered space when it was in use. So when you went in you were supposed to change the sign on the door to

match your gender. There were these generic men and women stick-
ers and a magnet and you were to place the magnet on the gendered
person that you were so that other people who identified with the
sign you've placed the magnet on could go in while you were [in
there] going [excreting]. . . . I think it caused people a lot of stress
because if you forgot—god forbid you forgot!—to put the magnet on
the person on the door, or someone ignored the magnet person, then
someone of another gender might come into the bathroom while you
were in it. This definitely caused some people a lot of anxiety. I never
really cared because the bathroom had two stalls, it was like, whatever, there are two stalls inside this bathroom, the bathroom itself does
not need to be gendered.

Basic bodily care in this genderqueer toilet was already problematic
because of the added threat of accidental transgression and the negotiation of the different sets of gendered rules. For example, to simply
urinate in the 'correct' (i.e., non-threatening) way, people were expected to not only maintain the rules of the TIO, they also had to
consider that people of the opposite gender would use the space, and if
they were not careful, at the same time as them. This situation means
an already heavily coded and charged space is escalated to an extent
where people seem to forget the need to care for one's body.

Miriam told me this story of her co-worker, who, when trying to take
care of his bodily needs, made many enemies:

I had a co-worker who I think had some serious medical issues and
he would really make the bathroom smell really badly and it would
carry through the hall to his office and our offices, and nobody
wanted to go into the bathroom for the little while after he used it
and it was just awful. It seemed like he was really sick, but no one
would be like 'hey man, get some air freshener' or something. It was
very disruptive to my work environment because everyone [in the
office] was talking about it all of the time. They hated him for it.
They hated him. Everyone was really frustrated, but no one would
ever say anything to him about it. It happened every day. If he wasn't
in his office someone would come to the area where our offices were
and be like, 'did he just go to the bathroom?' And ask one of us
where he was! I didn't really care, I would say 'yes or no' but it was
just awful—the culture of that space during that time—and I would
just go to another floor and use the bathroom there to try to avoid
the situation.

This example of how transgressing the TIO can be so disruptive clearly
shows how people's bodies are open and interconnected in three ways.
First, there is Miriam's male co-worker whose bodily care was so highly

disruptive to the point of his co-workers 'hate[ing] him' for his need to defecate. Apparently there was almost no respect for this man's need to take care of his body, save for Miriam's concern. Second, the policing and discussion of his bodily care practices by everyone in the office except for him exposes the social interconnectedness of people and demonstrates how an extremely negative and damaging environment is created based on one's bodily needs. It is almost as though the rules of the TIO are being enacted on his behalf outside of the toilet space. As if he did not feel enough FASE about his body, and therefore people inscribed it onto his body for him—reflecting the utterly social nature of the TIO, which aims to keeps bodies radically atomised through a valuing of those very emotions (FASE) which make people feel unfit for social life. Third, this gossiping and policing culture made Miriam change her own habits and begin using a different space for some time. Taken together, this is an example of material-discursivity in action.

Taking this into consideration we can better understand the pressure Miriam felt when she had her own need to care for her body in a similar way. She explained:

> When it came to my co-worker, I was always trying to find a can of air freshener or something to keep in that bathroom, or I would call facilities and ask them to leave one in the bathroom so people could use it and I would use it when I thought I made a mess, a smell in the bathroom. Because there was a time that, when I was in that office, I was on a medication that made me poop a lot and really smelly and it wasn't something I could control—if I had to poop, I couldn't stop it and it would verge on being diarrhoea and I ended up stopping that medication because I couldn't deal with the gastro side effects from it. Thought I was on it for eight or nine months.
> [During that time were you concerned about the office culture?]
> Totally. Oh of course. That's why I was so careful about it—I would even lock the door even though it was a two-stalled bathroom, so nobody could come in and that caused drama too! Someone would always be like, 'this isn't supposed to be locked!' I was just like 'why the fuck not?!' Yes, there are two stalls but maybe, sometimes you want that privacy. I just wanted to be like, 'look I'm sparing you these awful, awful stomach problems that I'm experiencing!' That experience deteriorated my quality of life I think—being on that medication and a big part of that was the work-bathroom culture. . . . I do think that that experience did make me more comfortable with pooping in public because I couldn't not. I couldn't hold it back. I know you're not supposed to do that [hold in your excrement], but there are ways that you can do that to yourself if you're out for a little while or don't want to poop at work. I don't do that often and I do

poop regularly at work now, whereas before I was on that medication
I would avoid it.

Miriam's bout with the poor body-toileting culture at her place of work
was at once anxiety provoking and in some senses liberating. While the
medication made her ill and, as already demonstrated, her co-workers
seemed to have ownership issues over the space and what people could
and could not do in it, Miriam gained a more open and comfortable
relationship with her bodily needs. She did this through the practices of
caring for her body, despite protest and malice from her co-workers.

While her initial inclination may have been to stop herself from
defecating at work because of the extremely negative culture of the
space—an example where language (i.e., discourse) itself domesticates,
retains, sets boundaries onto bodies (i.e., materiality)—when she physically could not do that and allowed her body to openly, without restriction, experience what it needed to, she was able to overcome much of
her FASE, and now she regularly defecates at work. Considering the
utterly damaging social culture at her workplace surrounding the open
expression of intimate bodily needs, the change in Miriam's habits is no
small feat. Here her co-workers attempted to straighten, order, and
impose stability on the already always unstable-gendered space,
through the policing of bodies open in excretion. This openness is never
singular but rather underscores the openness of all bodies. By unapolo-
getically taking care of her bodily needs, Miriam undermines her co-
worker's *homo clausus* efforts to remain individual and unaffected by
other people's bodily openness.

CONCLUSION

This chapter has explored the various ways people, in their mundane
daily lives, are interconnected with others in public toileting practices
of care. Those caring practices take many forms; they are sometimes
publicly visible and at other times happen behind closed doors, but are
always already implicating other bodies in and around the space of care.
These expressions of bodily openness and interdependency are examples of people as *homines aperti*. Even when such practices are resisted
or challenged in an effort to maintain *homo clausus* individuality and to
impose FASE onto others, they continue to highlight the interconnectedness of people. While fear, anxiety, shame, and embarrassment continue to play a large role in the coalescing of *homines aperti* embodiment, practices of care point to the fissures inherent to *homo clausus*

individual identity, exposing where greater intervention in one's own sensory-embodiment can take place.

NOTES

The opening quote is from Norbert Elias, *What Is Sociology?* (New York: Columbia University Press, 1978), 137.

1. Arthur Frank, *The Wounded Storyteller: Body, Illness, and Ethics* (Chicago: University of Chicago Press, 1995), 48.

2. Elias, *What Is Sociology?*, 135.

3. Elias, *What Is Sociology?*, 135.

4. Elias, *What is Sociology?*, 136.

5. While I don't have the space to develop it, it is important to mention that this is also clearly an example of ageism, which operates through social policing to keep older 'abject' bodies out of public life.

6. Elias, *What Is Sociology?*, 137.

7. Julia Twigg, 'Carework as a Form of Bodywork', *Ageing and Society*, 20, no. 4 (2000): 408.

8. Twigg, 'Carework', 389.

9. While I don't reference it here, there are some interesting and direct parallels regarding self and other in this section/chapter and Derrida's work on hospitality. See, e.g., Derrida (2005) and Derrida and Dufourmantelle (2000).

10. Twigg, 'Carework', 394.

11. Twigg, 'Carework', 406–7.

12. Twigg, 'Carework', 408.

13. Twigg, 'Carework', 409.

14. While this space is generally referred to as 'gender neutral' I think 'genderqueer' is more appropriate since people were expected to actually change the sign on the door based on one's own gender. Gender neutral spaces are not gendered, whereas this space has a gender that was constantly in flux.

MODERN ADVENTURES AT SEA

Peter Gizzi

Say it then or
sing it out.
These voyages, waves.
The bluing of all I see.
Sing it with a harp
or tambourine.
With a drum and fiddle.
These notes and its staff,
the lines' tracery
blooming horizon.
These figures insisting.
Their laws. I embrace
accident. I accidentally
become a self in sun
in the middle of day.
Where are you? Cloud,
what shadow speaks
for me. I wonder.
Is there an end
to plastic. Is
yesterday the new
tomorrow. And
is that a future?
Do we get to
touch it and be
content here
before we go.
That the signs
won't remain
untranslatable

in the end.
And that I may
learn this language
say with a dolphin,
a dog, a cattle herder
and slaughterhouse,
a lumberyard
and big redwood.
That someone
could say to the crude.
Stay there. And
don't be drawn
into this tired game.
I wonder if the poem
gets tired. If
the song is worn
like sea glass.
I wonder if I am
up to this light.
These ideas of order
and all I feel
walking down
the avenue.
I see the sap
weeping on the cars.
See the wrack
about my feet.
Its state of decay.
To see that decay
as the best of all
worlds before me.
It's transformation
not transportation
that's needed. Here.
It's embracing
the soft matter
running my engine.
My guff. And fright.
This piss and
vinegar. And tears.
That I won't
commit violence
to myself in mind.
Or to others
with cruel words.
That I may break

this chain-link ism
of bigger than
smaller that why
feel bigger than
anyone feeling smaller.
Can I transform
this body
I steward. This
my biomass.
My accident.
When lost at sea
I found a voice,
alive and cresting,
crashing, falling
and rising. To drift,
digress, to dream
of the voice. Its
grain. To feel
its vibrations. Pitch.
Its plural noises.
To be upheld
in it, to love.
Whose book lying
on that table?
And where does
the voice
come from?
What life was attached
to its life,
to its feint,
its gift of sight.
To understand
oneself. With-
out oneself.
How to live.
What to do.

7

CORPUS INFINITUM AND THE MATERIALITY OF POSSIBILITY

This chapter explores various and varied experiences of public toileting practices which further challenge the rules of the *homo clausus* triadic intra-action order (TIO) that, when adhered to, attempts to maintain the monadic experience of the self-body as stable, controllable, and closed. In this chapter I show how the rules of the TIO are not unchallengeable but rather highly contestable and instable. The stories presented in this chapter coalesce around themes of play, pleasure, and possibility—terms I use loosely and will explicate in further detail below—and expose how one's bodily being in and of the world can shift from rigid habit to open, boundless becoming. I conceive of these shifts as happening in momentary thresholds where habit is released; they may only last for a few minutes or even seconds, but they are vital for becoming-other. These moments are what the material-discourse of new ways of being are made of—they are the stuff of difference, of change. When thresholds are opened and explored they enliven more thresholds that, over time, can unhinge habits and entangle new potentials in the everyday. That is to say, habits are not simply broken and removed from someone's way of being, but rather weakened over time. New ways of being are found, explored, and allowed to become, allowed to have an effect—they are not instituted or simply adopted. Habits, while useful in daily life, focus on the effect of stability and sameness while denying possibilities of/for difference. Habits are required for thinking we are stable selves (see, e.g., Butler 1993), and they attempt to remove the thresholds of difference through perpetuating fragmented self-experience. Alternatively, the nature of becoming-other is directed towards recognising thresholds as opportunities to become-other. Practices of bodily becoming move beyond lived habits and provide new opportunities for being and exploring one's self as inherently social and spatial—that is, always already entangled in the dynamics, actions, and understandings of the people and places of an experience. This is a shift from a *doing-focused* self-body to a *becoming* embodiment characteristic of *corpus infinitum*.

Practices of play, pleasure, and possibility point to the potential of situating oneself outside of the *homo clausus/homines aperti* dialectic of understanding, insofar as they highlight the nascent dissolution of ima-gined bodily borders of the individual self and allow for a continuity of experience not dictated by habit and representation. Whereas *homo clausus* is 'stabilised' through processes of territorialization of the body and *homini aperti* is 'stabilised' through a constant state of de-territori-alization of the body, *corpus infinitum* highlights the spaces of entangle-ment between stability and fragmentation, where cohesive difference and creative, curious openness are fostered. Thus *corpus infinitum* is always already before and beyond de-territorialization.

Rather than disregarding these stories as mere anomaly or unimpor-tant leisure activity, I use them in order to further expose the bound-less, unfinished nature of the social-spatial-bodily-self. While many of the examples used throughout this chapter take place in contexts of 'leisure' (e.g., bars, concerts, general socialising), the practices I have chosen to focus on need to be understood as decidedly different from and even outside of work and leisure activities in order to draw out the intra-related nature of the body-self-society onto-epistemology. As Ross, speaking about French poet Rimbaud's relationship to work and the social space of the Paris commune explains, 'The refusal of work is not an absence of activity, nor, obviously, is it leisure since leisure reinforces the work model by existing only with reference to work: it is a qualitatively different activity, often very frenetic, and above all comba-tive.'[1] The material-discursive practices outlined below, on many levels and in many forms, express a bodily desire for openness and connec-tion, for a becoming beyond one's self that is pointedly a refusal to *do the work* central to the sustenance of the *homo clausus* triadic intra-action order. My reading of this refusal to do the work of *homo clausus* embodiment can also be understood as an effort to further challenge 'the widespread notion that there exists a social production of reality on the one hand, and a desiring production that is mere fantasy on the other.'[2] Therefore, I aim to show how frameworks of onto-epistemolo-gy, when applied to daily experiences of the desiring body in public toilets, can usefully problematise the separation of the individual body and social life, and ultimately point to new practices of sensory-embodiment that can be understood as combative to social norms and potentially liberating from individual monadic identity.

The empirical data presented in this chapter have been divided into three categories that are neither discrete nor concrete, but rather open, malleable, and entangled. In some instances the practices themselves express a direct relationship to the category they have been placed in— e.g., the use of pleasure to better enable urination or instances of lesbian sex are both clearly practices of pleasure. In other instances there is a particular sentiment or sensory-embodiment (sensation, feeling) that is part and parcel of a practice but may not be immediately or overtly understood as such through a rational form of understanding. For example, conversations or interactions which establish an openness between friends, particularly when it comes to bodily practices, are often necessarily of a playful nature, often with jokes and laughing which help avoid individual shame and instead express embarrassment in a communal way; by laughing at oneself and one another, the members of the group overcome their personal FASE. While conversation about public toileting practices may not initially seem like a form of play, when we consider how it operates amongst friends and in opposition to *homo clausus* norms, it can be understood as such. Furthermore, while many of the examples used below are made possible through the consumption of alcohol (certainly many of the stories were only possible because of the need to urinate after consumption of alcohol), these instances should not be dismissed as outside of normal social intra-active practices or of any less value. On the contrary, the inclusion of alcohol in many of these stories crucially highlights how easily habit and bodily repression can begin to be released, and *is desired to be released*, exposing how willing people are to disregard the social norms of the *homo clausus* triadic intra-action order, in order to foster new forms of sensory-embodiment leading to social cohesion. I follow Rimbaud's sentiment again here, as Ross describes, for Rimbaud, 'intoxication is not a dulling, a numbness or impoverishment of sensation but rather an activity: too much sensation rather than too little.'[3]

Practices of play, pleasure, and possibility embrace and express bodily desire to become-beyond individual borders of the *homo clausus* self. The material-discursive practices elucidated in this chapter further highlight how and when everyday embodiment can be released from the rigid patterns of normative use and reconfigured for more expansive, dynamic, and collective bodily experiences. By giving attention to and recognising as personally and socially significant non-normative practices of play, pleasure, and possibility we can build a greater understanding of how and when the self-body is habitually hemmed into and condensed by *homo clausus* social roles, rules, and expectations and thus locate power in those practices which are not typically understood as useful, let alone powerful.

PLAY

Play can take many forms, though it is not necessarily readily recognised as a potential (let alone a generative potential) in daily adult life; it is generally associated with children and thus understood as childish.[4] Play in which adults partake is typically understood as 'leisure'—a way people use their time when they are not engaged in work—and is prescribed according to social norms. This means the immanent playfulness of everyday life can be systematically lost, ignored, or forgotten when *homo clausus* adulthood is reached. In this section, I highlight forms of play and playfulness that occur in daily life, as part of both 'leisure' and 'work' time-space, within the ambiguous, marginal regions of public toilets. In doing so I aim to show how people actively break the rules of the triadic intra-action order, forgoing fear, anxiety, shame, and embarrassment in a desire to foster greater connections with those around them by way of play.

Steve, a twenty-four-year-old straight man, shared two relevant stories with me regarding play and playfulness in toilet spaces. The first is related to his experiences of stage fright (inability to urinate in the presence of other men) and overcoming his shame and embarrassment in talking about it. He connects this experience to one from his childhood, when he began masturbating:

> Initially, it [talking about stage fright] was something that really bothered me and I guess the best comparison I have is like talking about sex and I guess masturbation is the thing that comes to mind, because I remember being a kid and when I started masturbating, and the idea of talking about that in public was like giving me the willies, and I remember being a kid and other kids joking about it and it making me feel so weird and now, obviously, I'm talking to you about it and I have no problem whatsoever, and it is sort of like, I would say it happens in an instant where you have some sort of common experience—it usually happens with another person, another friend, another guy of mine, where we're like, 'fuck yeah masturbation is awesome', or like, 'fuck yeah I have stage fright too', and from that day on it is something I'm totally comfortable talking about. I would say humour is a big thing for me as far as like being comfortable with things I might not be comfortable with in the first place. Sort of laughing about it definitely helps. . . . But then once—I actually remember—I went, I was at a music festival with like five friends

and we were all like drinking and walking around together and sort of were keeping the pack together, so we all ended up, maybe like three or four of us, going to the bathroom together and we all walked up to urinals and then we all came out and all of us were like, 'dude I didn't piss, did you?' And it was that moment where I was like okay, this is funny, we can talk about this, we all deal with it—so yeah, there was definitely a switch and now I'm completely comfortable with it; it doesn't bother me. I think it is just kind of funny.

Here, Steve explores how he uses the playfulness of humour to open new possibilities in his ability to connect with his own sensory-bodily experiences and with the experiences of his friends, allowing for more open, honest, and comfortable relationships beyond the confines of *homo clausus* masculinity. It is not insignificant that he situates the experience of connecting with his friends over the embarrassing experience of stage fright, both within terms of early childhood sexuality (i.e., pleasure) and the use of alcohol, as they are both forms of sensory-embodied becoming that have the potential to de-territorialize the boundaries and borders of the *homo clausus* self. Steve's ability to overcome his own bodily embarrassment through amusement and laughter enabled him to release FASE in intra-action with his friends, making space for greater sensory-embodied being.

Similarly, and before going onto Steve's second story, Kat, a twenty-six-year-old queer female, shared a story with me about her experience, also at a music festival. She says:

I was at a festival with my friends and we were given those SheWee things [which are supposed to enable women to pee standing up] and so a couple of my friends and I tried to use them at the same time in the port-a-loos and it totally didn't work. I mean, I just wee-ed all over myself, all over my jeans, and when I came out I just told my friends, 'I just wee-ed on my jeans' and they had too, so it was just hilarious. If the circumstances had been different and if we hadn't told each other, if we just tried to ignore it, it would have been totally embarrassing, it would have been different, but since we talked about it, it was fine; it was actually just really funny.

As evidenced in previous chapters, speaking about urination and defecation can be especially difficult for women, as it is particularly taboo. Kat and her friends, by being honest, open, and playful about their experience of using the SheWee, were able to use humour to assuage the potentially uncomfortable and territorializing nature of *homo clausus* shame and embarrassment, enabling new ways of being in and of the world and being with one another—that is, no longer having to feel embarrassed about bodily functions both individually and collectively.

In Steve's second story he again uses play and humour to connect with his friends in other ways in public toilets. He explains:

> I mean usually I know my friends and if it is okay to joke around with them, to, ya know, try to embarrass them, and so we'll purposely disrupt and break the rules of the space by doing or saying something to one another, where everyone else in there can hear or see, and it's interesting because I've been on both ends of that, where sometimes a more rambunctious friend of mine will do it [something embarrassing] to me and I'll sort of like be embarrassed, but usually we're on the same level and so we're in a public bathroom, we're having fun and it's not a big deal. . . . But I've been on both ends where I'm sort of embarrassing my friend a little bit or he's embarrassing me, ya know. But it is more that we'll say something outrageous or X-rated because it is suddenly quiet and everything we say, every other dude in there can hear . . . and it is just funny.

By pushing on the always already underlying fear, anxiety, shame, and embarrassment that one is expected to feel in these spaces, Steve and his friends are able to easily create playful situations which render those emotions much less powerful. Instead of FASE being experienced as scary and oppressive, in Steve's scenario they are positively engaged in order to lighten the atmosphere or break the tension of the space. What is produced is a fun and funny bonding experience, which Steve emphasises is always balanced, not a one-sided attack. He and his friends use these playful experiences to expand how they can interact with one another, with their bodies, in public. Additionally, it seems that since these acts are so public, that is, with the intention of the others in the space not only seeing or hearing, but also being affected by them, in a sort of collective experience of embarrassment, they point to the political potential of playful social interactions to diminish the power of FASE by purposely drawing it out.

These examples of play and playfulness highlight how normative social relations, which are aimed at maintaining the sensory-embodiment of monadic *homo clausus* individuality, can be reconfigured to lead to new ways of being beyond FASE. In the stories mentioned above, an individual, mundane daily activity is turned through collective mediation into a playful group experience based around failure. The failure and/or active denial to 'successfully' urinate according to the adult *homo clausus* rules of the triadic intra-action order, for example, opens the connective possibility of *corpora infinita*, as playful openings are used to cement friendship rather than result in shame and embarrassment that could disturb friendship. Similar to the media-focused examples Judith Halberstam explores in *The Queer Art of Failure*, these accounts make 'peace with the possibility that alternatives dwell in the murky waters of a counterintuitive, often impossibly dark and negative realm of critique and refusal.'[5] While one may rationally *think* that actively exploring, expressing, or exaggerating ingrained feelings of bodily fear, anxiety, shame, and embarrassment would only make those feelings worse, would only negatively intensify them, as shown above, in actual practice, intensifying or allowing those feelings to be intensified enables the de-territorialization of the *homo clausus* self and a move towards open, boundless, becoming as *corpus infinitum*. That is, a self-body open to and configured in ongoing social experiences, not a self-body rigidly defined and managed by learned habit. A similar phenomenon can be recognised in relation to practices of pleasure and intimacy.

PLEASURE AND INTIMACY

There is an abundance of (sexual) pleasure and intimacy occurring in public toilets. Since the spaces are so rigidly managed according to the triadic intra-action order, people are able to work that 'system' to their advantage in order to have pleasurable experiences. While these experiences are considered socially 'deviant', it is important to remember that they are generally only possible because of the social norms, rules, and codes that are so firmly established in those spaces to begin with. For example, because eye contact is so taboo in men's public toilets, men can use it as a way to quickly, discretely, and easily express the desire for a sexual encounter. Sex between men in public toilets has been well documented,[6] but very little has been written about women having sex in public toilets. While that is not the focus of this section, it is important to note that every lesbian and queer woman I interviewed had had multiple sexual encounters in public toilets. Lesbian and queer female sex happens just as much if not more than sex between men in public toilets. It is a norm. This is an interesting sociological finding, but it is not entirely of interest here because these practices do very little to challenge the *homo clausus* triadic intra-action order, since they generally rely on its maintenance for such transgressive behaviour; the behaviour is not 'deviant' or 'transgressive' without the norms which straighten and order the spaces. I note this here not to place a value judgement on such transgressive behaviour, I certainly do not think there is anything wrong with it (and generally believe 'deviant pleasure' in public can be powerful politically), but rather to point to how transgression and deviance of this sort does very little to challenge norms. As Bataille points out, 'There exists no prohibition that cannot be transgressed. Often the transgression is permitted, often it is even prescribed.'[7] Furthermore, I want to make clear that for the people who spoke openly to me about having sex in public toilets, the practice is not one of shame but rather one of excitement and exploration. That is to say, those people I interviewed do not engage in sexual acts in these spaces because there is 'nowhere else for them to go' (as goes the general, outdated trope for homosexuality), but rather because they offer a different way of being sexual and intimate which is not as readily accessible to heterosexual couples. This is a point I will develop further in the possibility section of this chapter. So while I could fill several pages with stories of sex in public toilets from my participants, it would do little to further my theory of experiencing the self-body as *corpus infinitum* (they would instead approximate *homines aperti* ways of being by capitalising on the rules of the TIO for their transgression). Instead, I focus here on experiences of pleasure and intimacy which explicitly challenge the norms of the triadic intra-action order; they are practices which *refuse to do the work* of maintaining a *homo clausus* self. These experiences work towards undoing habit and creating new ways of being in, of and with others in the world.

The first example comes from Josh, a twenty-nine-year-old straight man. His story takes place at a university where he was working. He explains:

> Just the other day when I was [on campus] doing some work I went into the toilet and two guys were having sex! It was late afternoon or early evening, the end of the day. I went to the bathroom with my headphones on so I wasn't really paying attention, and then when I was finishing up I took off my headphones and realised there were two guys engaged in some kind of sexual act in the next stall—I guess they hadn't realised I was in there either—because it was like as soon as I noticed them, they seem to know that I knew they were there—or maybe they just realised I was there too. Anyway, they realised I was there and realised that I could hear them and quickly redressed and left the bathroom. I just kind of sat and waited for them to be *totally gone*—I didn't want any chance of seeing them or knowing who they were [because I'm a lecturer]—and then I quickly left myself. I was really surprised about this whole thing. The experience really pulled me out of my drone routine.

Josh's story both confirms the norms of how people have sex in public toilets according to at least some of the norms of the triadic intra-action order—we can infer this by how quickly they left once they knew that

Josh noticed them—and also points to how such an experience can disrupt habit. The most interesting part of Josh's story is in the last line of the quote: 'The experience really pulled me out of my drone routine.' What is vital here is not that Josh caught these men in a 'deviant' act of public sex, but rather that the act caused Josh to *feel* differently; the experience caused him to not only take notice of his disconnected, habitual way of being, but also to feel surprised—a sensation of *becoming*—without any of the normative FASE. This rupture is where potential and possibility thrive because with awareness comes choice to change behaviour. While it may have been a one-time or fleeting experience for Josh, it is a small fissure in the mundane opened up through a connection of pleasure, where new ways of being and doing the self-body can be forged.

David, who is a thirty-six-year-old gay man, shared a story with me about reading poetry in glory holes, which are small openings in the walls or partitions separating cubicles in public toilets and buddy booths (small rooms where people go to watch porn). These openings are generally created specifically for sexual encounters, which could be anything from just watching someone urinate or masturbate to directly, physically receiving pleasure from the person on the other side of the wall. He says:

> [These spaces] have always fascinated me. A place where one can walk in off the street, fulfil their fantasy and then leave. I liked the accessibility it offered—anyone . . . could go in. But my fantasy wasn't so much getting off [having an orgasm] with guys whose full body I can't touch, rather, I wanted to connect with them personally in a context that *allowed* for instant sex. And so I wrote a poem, which perhaps gives me at my most vulnerable, and I read it through glory holes, [these were] mostly narrow rectangles, and performed it a few times. . . . There wasn't much interest in what I was doing, as even *talking* generally betrays the codes of these places.

Here David was not only interested in direct explorations of pleasure, but also in using the context of pleasure to foster connections, intimacy, in ways that are not necessarily expected, desired, or welcomed. What is of most interest here is his desire to connect personally in a space that allowed for public sex, but not by way of public sex—that is, to use the physical openness people allow in public sex in an attempt to connect with them at a different level, through personal emotion and vulnerability expressed through poetry. Just as Josh experienced in the previous story, David's intention was to interrupt people's habits by interjecting a form of personal intimacy and vulnerability that is not usually present or welcomed in situations of anonymous sex because it was neither explicitly sexual nor anonymous. David purposely engaged in an activity—reading his extremely personal poetry that made him feel emotionally vulnerable—within a potentially sexual context in order to explore the possibilities inherent in a daily social situation. This sort of practice, which challenges the norms of the space and, instead of doing the work of the *homo clausus* self, purposefully explores feelings of weakness, openness, and exposure—normally understood as threatening and degrading to *homo clausus* masculinity—exposes how the power of FASE can be significantly weakened to foster different, more comfortable, cohesive, and powerful ways of being. Additionally, it is not necessarily important that he felt there was not much interest in what he was doing; we have no way of knowing how his actions impacted those he could not see. His unusual, poetical form of interacting with glory holes—an already 'deviant' spatial feature of public toilets but one that is created within the bounds of *homo clausus* being—points to the possibilities of *corpus infinitum*, of open bodies and social relations in the everyday at a subtle and nuanced level.

The next example of the power of pleasure and intimacy is from Lana, who is a thirty-year-old queer woman. Here she is speaking about using public toilets with people she has been in relationships with. She says:

> Yeah we [my girlfriend and I] have used a public toilet together and I remember because it's very noticeable, and it's not very often that we're in a place where we would need to do that together—I guess most recently we [were at a concert] and we were there [in the toilet] and it's, I feel like she is uncomfortable with it so I don't express anything toward her when I'm in there with her. That is specific to this relationship—it is a little uncomfortable, because I know that she is uncomfortable. But in other relationships, I'd like stick my head under or over the stall, and fool around with them [my girlfriend], like tease them and I would put my hand under the stall, I was totally comfortable and playful about it. But I think that she [my current girlfriend] is more uncomfortable, so I don't do anything.

Not wholly unlike Steve's joking play with his friends, here Lana explores how intimacy and playfulness was expressed in previous relationships compared to her current relationship. Since her current girlfriend is uncomfortable in toilet spaces and with toileting behaviours generally (something we spoke about at length), the potential for intimacy is precluded and Lana herself feels uncomfortable. What is interesting here is the variation of Lana's behaviours based on the person with whom she is in a relationship. From playful expressions of intimacy to absolute rigidity, her story shows the differences in the social potential based on how her partner feels about her self-body in public toilets. When with a more open, comfortable partner Lana felt it was possible to express a queer form of intimacy in a heteronormative social space where such an expression between partners is typically precluded.

Lastly is another story from Steve, again about his experience with stage fright and a pleasurable body technique. He says:

> Sometimes, when I do have stage fright I'll try rubbing the tip of my dick, almost in a sort of like a sexual arousal sort of way, but it can also help me pee too, I mean it's not to arouse me, it's to get the pee out. Sometimes after sex when I'm still sort of aroused, I'll rub my dick a little and it's like magic genie, I'll pee. So I'll sometimes use that technique when I'm having stage fright and sometimes that works, but that is also even more awkward if some dude looks over and I'm rubbing the tip of my dick. Wow, I'm very comfortable talking about this stuff right now.

There are a few points in this story worth exploring: the open blurring of a mundane, utilitarian bodily act of excretion with pleasure; the openly public practice of that act (where someone may see); and Steve's surprise in how comfortable he was at this moment in our interview when speaking about this body technique. I'll deal with these in turn. First, since pleasure in the context of *homo clausus* public toileting practices material-discursively challenges social norms, it is important that Steve uses a pleasurable experience to help him cope with the *homo clausus* pressures of masculinity. This is where sensory-embodiment beyond the normative self serves one's daily experience of being and having a body—the use of pleasure to combat rationally embodied pressure points to the nature of possibility inherent to *corpus infinitum*. It is an example of what is possible when someone overcomes the rigid idea of a *homo clausus* self by purposely engaging in a risky act—an act that could surely be read as non-heteronormative and thus suspect and threatening according to the triadic intra-action order. This leads on to the second point, of publicness. It is remarkable that Steve engages in this material-discursive act openly, enabling a reordering or condensation of the public-private dynamic, as it is an act which may seem counterintuitive for someone who experiences stage fright. The logic being that if someone experiences the potentially de-masculinising phenomenon of stage fright and, even worse, is caught by other men while it is happening, the effects on the self can be personally damaging, as the TIO works by individuals engaging in hetero-masculine material-discourse. Nevertheless, Steve, ignoring (or in direct response to) the social pressure to maintain *homo clausus* masculinity, pleasures himself in order to help him urinate. While he recognises this may lead to an awkward experience if someone sees him rubbing his penis, he does not at all seem concerned about that awkwardness escalating or having damaging effects on him. It is merely the reality of the situation as he is openly breaking the rules of the *homo clausus* triadic intra-action order and not allowing fear, anxiety, shame, or embarrassment to curb his embodiment. Thirdly of note is Steve's open expression of this practice to me; this is another intra-active material-discursive expression of *corpus infinitum*, an opening used for expression instead of ingrained or expected embarrassment. In articulating this body technique with me Steve moves beyond the FASE that may normally constrict his ability to speak about his sensory-embodied experience and, in an act of becoming, recognises his ability to surprise himself. This retro-aware experience of becoming animates the final section of the chapter where I focus on two individuals' experiences of making possibilities anew for everyday sensory-embodiment.

POSSIBILITY

In the third and final section of this chapter I orchestrate an extended exploration of the potential to shift from *homo clausus* doing to *corpora infinita* becoming. While it is impossible to capture sensory-embodied becoming in words *as it occurs*, it is vital to understand (albeit retroactively) where, when, and how those experiences materialise and come to matter—that is, how they happen and are made 'sense' of. To do so I focus on the sensory-embodied experiences of two individuals in this section. First is a story from Joseph about the mundane turning into different forms of charged intimacy and the second is from Zevi, who engages in a more sustained, fleshed-out, and encompassing exploration of the possibilities of becoming in relation to his self-body identity.

Joseph, a twenty-six-year-old queer man, shared this story with me about an outdoor concert he attended on New York City's Governor's Island one summer night. He explains:

> We were in the VIP area with its own port-a-potty toilet . . . and a bunch of us were waiting in line. There were maybe eight or nine people in line at that point and it was very dark outside, but we had all been outside all night so we could all see . . . everyone in this line was kind of chatting, not necessarily with each other but with the people they were waiting with and there was some acknowledgement that this many people waiting for one toilet to use in the dark was going to take a long time. . . . So, someone, or somehow the conversation started about 'well, why don't the guys just go and pee some-

where else?' because despite this being a big public concert, we were in this sort of secluded area. . . . So, it seemed at first like a joke, like 'haha, wouldn't that be funny?' and then I was just sort of like, 'well, that is actually a good idea, why don't we just do that?' And there was some discussion about it . . . so it wasn't like this instantaneous decision. There was sort of some banter in the line about it, it was kind of a humorous, light conversation and a lot of people were like, 'Yeah! That's a good idea', but then like, didn't do it. So then at a certain point, I asked [my female friend her opinion on it and she was like], 'yeah that's a great idea' and I was like, 'okay!' . . . And I thought there would be this like exodus of people because the line was fairly split between men and women. . . . So when I left the line I thought other people would leave and just find their own space to pee in . . .

So I left the line and this guy kind of left with me, but I don't remember explicitly talking to him about it, I think he was just part of the group chatter. . . . But then as soon as I left the line he very clearly left the line too and walked next to me, and I was like, 'Oh! he's walking with me', and then he maybe introduced himself to me and all of a sudden there was this sort of, I kind of realised it was just the two of us that were going, so it kind of took on a different air at that moment, and I was like, 'oh, this is different!' and then it be- came clear that he was walking with me. . . . It was this sort of weird instant negotiation or bonding of just like, 'oh well, you and me are the ones who are going to pee so let's do it!'

So we both walked over to this spot, not far from where the toilet was and there was like a bush and building and a little ditch area, and we basically stood right next to each other and we were talking; there was a continuous conversation, and I had a moment where I wasn't sure exactly if I needed to pee or if I was going to be able to pee in that situation, so there was a little hesitation, but then we both basi- cally got out our penises and started peeing, and still talking, and there was this sort of light, playful, tone. . . . I think once I realised that I was going to be able to pee it was fine. I had a moment where I thought I wasn't going to be able to, and then I was like yeah, I actually do need to pee. And then he told me to make sure I didn't pee in his drink, and I said I wouldn't. And then there was clearly a moment where he looked at my penis and I looked at his and there was this thing where it immediately became more overtly sexual, whereas it was just implied somewhat before. It was flirtatious and somewhat aggressive or direct, but I wouldn't say it felt like *SEXU- AL*. There was this sort of innocent, kind of summer whatever, care- less attitude. I don't know how to explain it, it was just like, 'Oh we're outside and we're doing this thing together', and I realised, 'Oh you're flirting with me, that's why you're coming with me'. Then I was like, well this is a pretty direct kind of flirting. So when we both finished peeing, we turned to face each other and I think I said, 'I feel like I should kiss you now' so I gave him a little kiss, and there was sort of this moment of something, and that was it, and we just walked away, and I just felt giddy from the whole thing because it was fun, it was like a surprise!

. . . It was more of an enjoyable, playful experience, than a functional one. Yes, I was peeing, but it was like peeing as a form of flirtation in a way. Ya know, it was something we were doing together. It was more fun, more exciting; there was more possibility that something could happen than going alone. It offered some openness; I think if either one of us had pushed a little more, something [sexual] could have happened right then. . . . There was an abruptness to it. . . . It was a different kind of intimacy. But it also felt very common—it was just like this acknowledgement—'well as long as we're doing this, we might as well do it together and check each other out'; there was an inherent opportunity to that moment, which is not necessarily as foregrounded in other peeing situations. Where it is just very self-focused and just get in and get out. It didn't feel particularly rushed. I think once we both felt comfortable there and peeing, it was like, okay, this is fun.

Put most simply, Joseph's experience is one of possibility. In literately leaving the straight, narrow path of the *homo clausus* intra-action order (upon leaving the line for the port-a-potty, after affirmation from a woman) and creating a new, unordered space for the mundane (in the unusual open-air, outdoor setting), Joseph vitally took advantage of a fissure in daily habit and dis-embodiment. In doing so he opened his experience to any number of possibilities and, at least momentarily, shed his ingrained bodily fear, anxiety, shame, and embarrassment, allowing an opening for new ways of being in and of the world. This enabled him to explore a new kind of intimacy, which was not necessarily sexual but, rather immanent and becoming in the moment. This is decidedly different from gay cruising or cottaging insofar as there was no intention or expectation on Joseph's behalf; the encounter was not exclusively or conventionally sexual, but rather flirtatious, surprising, with an intimate playful friendliness. Joseph's exciting and endearing story is one of potential. He and the man who joined him in urinating that summer night purposely veered away from the *homo clausus* intra-action order and explored a new, innocent sensory-embodiment within the mundane. This powerful, unnecessary indulgence of possibility transforms daily habit into open exploration; it drains the mundane of seriousness, of rigidity and infuses it with playful possibility and a nascent power of becoming beyond the self. No longer a singular, strict act, urinating in this story highlights how bodies can (be)come together, allowing for a new sensory-embodiment.

Zevi, a twenty-three-year-old queer man, whose story I'll flesh out in detail below, cruises public toilets for sexual and intimate encounters. While generally the stories and practices regarding sex in public toilets that I gleaned through my research do very little to challenge the *homo clausus* triadic intra-action order, Zevi's are different in one fundamental way: Even while he utilises the rules of the triadic intra-action order to enable his explorations and adventures, he uses them to get before and beyond the *homo clausus* embodiment that the rules are seeking to maintain. His practice, while focused on extracting new possibilities out of his sense of self, is aware and considered. He explains:

> What I have been doing, it is site specific, [I've] been going to Grand Central Station . . . because it is totally open to the public, and it is this huge place where anybody from anywhere can just walk in and be there. I'm really fascinated with A, the different, people from all these different places meeting in this one place that is also very private, and [B] it has this, I don't know if it is notorious, but it has this history of being this really cruise-heavy restroom, this legacy. I was very fascinated with that and seeing if it was still going on and if it was, what did it look like, how has it changed, ya know, everything about it. So I go and hang out for about, like a couple hours at a time, or until I got involved with someone . . . but yeah usually I just observe.

As a young gay man, new to New York City, Zevi was interested in connecting with a history, a material 'legacy' from those queers who came before him. He is interested in the context and chooses to place himself in it to explore parts of himself. He says a bit more about this:

> When I moved to New York, [cruising] Grand Central was like something I read about and also heard about from family members. I remember once an uncle of mine, we were at a big family gathering, like Thanksgiving, and something was mentioned, like how he had taken the train into Grand Central and had used the restrooms and was like, 'oh you gotta be careful, you never know who or what is going to be standing next you', something along the lines of like it being a highly sexualized place, and I was like, 'Ohhh . . .'. So it was in my consciousness and was something I was always curious about but never went to for a while, until like a year ago. And then I was like I'm just going to go and do it and see!

While cruising for gay sex in public toilets has been a part of cultural lore in the West for decades, for the younger generation of gays and queers, those of and around Zevi's age, it is something that many are surprised to hear still occurs, especially as a practice of a younger generation who have gay and non-gay bars, clubs, and pubs available to them. This is a sentiment both captured in my interviews (some young men found even the suggestion of this to be appalling) and strongly expressed by Zevi regarding his own social group and the social scenes he is a part of. Zevi's hesitation and curiosity about it are a case in point, but he also had this to say:

> There are a lot of people who feel the same as me or think it is interesting. And then there are some people who think it's odd, or are a little thrown off by it. Most of my friends think it is really amusing and awesome that I'm doing this. Some friends of mine didn't even know it existed anymore. They were totally thinking it was like a way of the past, and doesn't happen anymore. It is so interesting to me that so many gay or queer men don't think it happens anymore. . . . Yeah, another piece that was interesting to me is that not a lot of my queer friends were talking about it, and if they were, it was talked about as something that was a thing of the past that isn't done anymore, or that it's taboo, but really, it is really really active.

This is emblematic of the ways that material-discourse around gay rights and access to space circulates and how sex and sexuality continues to be socially managed to remain in the private sphere. As Zevi explains:

> I don't know where the disconnect was. Like this whole theory of queer or gay men having sex in public toilets because everywhere else was so policed and now this idea that queer or gay people aren't as policed, and we have more options and don't need those spaces anymore—which is not true. It is more exciting to be in these spaces.

For many young gay and queer people, including some I interviewed, the idea of having sex in public toilets is not only disgusting, but totally 'unnecessary'. The logic being that since gay people no longer 'have to resort' to these marginal spaces to satisfy their sexual needs, because they have gained at least enough social 'respect' to have sex like *homo clausus* heterosexuals in their own private homes, they no longer need public toilets for such transgressive acts. Even as gay and queer sex continues to happen in these spaces, the material-discourse surrounding it has changed enough to make younger generations (in these dense urban centres) feel unsure and curious at best and disgusted at worst by the continued existence of such practices. The *homo clausus* ways of being have been adapted through material-discursive practices to keep these sexually charged spaces increasingly non-sexualised in mundane daily use. This way of being precludes a series of possibilities which were once open to gays and queers, albeit due to marginality. Sex and sexuality as different ways of being in and of the world has lost a potentially powerful possibility in this normalising process.

Still, these acts, and Zevi's exploration of them, tap into the possibilities of an embodiment which is not hemmed in by *homo clausus* ways of being, even if only for the brief time of the encounter. As Zevi explains:

> I don't know what the people who are cruising in these spaces, what their sexuality is. It isn't important how people identify, it is just what is possible in that moment. Married men have come on to me there, usually gazing down at the floor, and there will be two hand dryers next to each other and we'll be drying our hands and they'll like really secretly brush my hand with theirs in this way that is obviously intentional and then we'll exchange eye contact. Yeah, it is really exciting. For me the married thing makes it even more so; it adds a whole other layer of, yeah, deviousness, for sure. The whole thing for me is really subversive. A lot of what I'm most attracted to about it are the possibilities. . . . An elderly Hasidic man also hit on me in Grand Central station and that didn't go anywhere but that was very interesting for me. I am a practicing Jew, but definitely not orthodox, but I mean, it was really interesting to me because I'm, I don't know, I'm sure like . . . if we were in his community, would he be doing that? And my response would be 'no, I'm pretty sure no, he wouldn't be doing that'. But I guess secular, out of the community space, his real desires, or like him tapping into his real desires, it is really interesting.

Even while some possibilities are only made possible because of existing social patterns which restrict and constrict how people can be, the crucial point is that people are *actively choosing* to allow themselves to be differently embodied beyond those self-socially imposed limitations. The power of possibility lies in the confrontation of fear, anxiety, shame, and embarrassment and going beyond that threshold of de-territorialization once again into comfort and grounded sensory-embodiment.

To further explore those possibilities of de-territorialization and becoming-other, Zevi has gathered his experiences and insights of cruising public toilets, of interacting playfully and pleasurably, into a performance piece that I saw performed during the course of my research, nearly a year after our initial interview. While largely movement and dance based, there are some monologues in the show. Two portions of greatest interest to this project flesh out Zevi's experiences of desire, exploration, and legacy. This section of text closes the scene that sets up the context for the show and his (re)telling of why and how he came to cruise public toilets. His recitation explains:

> And I wanted it, I wanted it all. I wanted the risks and I wanted the legacy, to see what changed. I wanted to see who's there and to see who isn't. I wanted to find the other world, I wanted to watch. I wanted new ways to experience my body, I wanted to know what my body could do. I wanted to just walk in and be there and not know where I was going but know that where I land is where I need to be. I wanted the fun. I wanted the journey, the chase, the dance. I wanted the secrets and I wanted the truth. I wanted it all. Tell me what you want, what you really really want . . .
>   A friend once said to me, '[. . .] you really are an old soul, except if you had been born two decades earlier, you'd probably be dead.'
>   . . . I go to these places, these bathrooms, to feel alive. When the married father was on his knees looking up into my eyes, when my grandfather died with me by his bedside, when my stomach dropped upon finding out my health had been put at risk by a lie . . . this is being alive.[8]

Zevi's is a story of living beyond the *homo clausus* self. He projects himself into the past and the future through intensely charged and emotional threshold experiences of becoming. In the monologue that closes the show, he is an archaeologist on a mission. He says:

> I am an archaeologist and I am on a mission.
>   A mission of exploration, walking narrow paths of the few and the foolish, amidst landslides and fault lines. The ground could give way, the earth could open up, the seas could part: I could reach the core.
>   Lookin' for tracks, turning over leaves, evidence of what might have been, of who might have lived. Lookin' for bones, a skeleton, tombstones, a butterfly trapped in time. . . . They say there's a secret inside of every stone, and that's why stones combust. And me? I'm just lookin' for those secrets . . .

Through his experiences of cruising public toilets, and the writing, staging, and performance of his show, Zevi becomes sensorially embodied in new ways beyond a stable sense of self. His experiences are de-territorializing and connective, aware and purposeful. As *corpus infinitum* he consciously releases his being in and of the world from *homo clausus* into *becoming beyond* rigid individual habit. In performing his show, he also affords others, that is, the audience, experiences of becoming, of going on a journey with him through differential experiences.

CONCLUSION

Throughout this chapter I have aimed to expose how practices of play, pleasure, and possibility are inherently powerful in and contribute to the de-territorialization of *homo clausus* sensory-embodiment. These practices are excessive, unnecessary, and often considered childish, in relation to dominant norms, and they powerfully enable people to feel surprised, excited, and different in their bodies. Fear, anxiety, shame, and embarrassment, in these examples, has been ignored, pushed, and productively utilised for the opposite of what it is learned for: the maintenance of the stable, controlled, bodily boundaries of *homo clausus*. Here instead, bodies are becoming beyond the individual self into new connections and intra-actions with and through other bodies. Sensory-embodiment, not rational habit, is the focus in these experiences and the way through threshold becomings into new ways of being. Practices of play, pleasure, and possibility seize the fissures inherent to *homo clausus* (in)stability and twist and push them in and through to openings that can be explored rather than anxiously closed. This is the potential of a radical politics of *corpus infinitum*.

NOTES

1. Kristin Ross, *The Emergence of Social Space: Rimbaud and the Paris Commune* (London: Verso, 2008), 59.
2. Ross, *The Emergence of Social Space*, 76.
3. Ross, *The Emergence of Social Space*, 112.
4. There are various viewpoints on the topic of adult work and play. See for example, Freud 1991, Huizinga 2003, Neff 1985.
5. Judith Jack Halberstam, *The Queer Art of Failure* (Durham, NC: Duke University Press, 2011), 2.
6. See Humphreys 1970.
7. Georges Bataille, *Eroticism: Death and Sensuality*, trans. Mary Dalwood (San Francisco: City Lights Books, 1986), 63.
8. The author has given me full permission to cite this and the following excerpts (from an unpublished script). I have not referenced them in the traditional manner in order to maintain anonymity.

IV

Entangling Ethics

Conclusion

TOWARDS A NEW ETHICS OF BEING

For the purposes of her argument in *Space, Time, and Perversion: Essays on the Politics of Bodies*, Elizabeth Grosz (1995) sketches two broad categories of feminist theory: first are those that take 'women' and 'the feminine' as their focus, and second are those that take patriarchal systems as their point of departure. She explains that the first type

> aims to include women in those domains where they have been hitherto absent. It aspires to an ideal of a knowledge adequate to the analysis or representation of women and their interests, and exhibits varying degrees of critical distance from the male mainstream. What distinguishes this group from the second are its interests in focusing on women or femininity as *knowable objects*.[1]

While this first type of feminist theory, which accepts the 'basic precepts and indeed implicit values governing mainstream knowledges and disciplines (or interdisciplines)', has been extremely fruitful, it remains inculcated in patriarchal systems and ways of life and is thus subject to the same 'crisis of reason' that mainstream knowledges are.[2] That is to say,

> Where feminist theory questions mainstream knowledges either to augment them; to replace them with competing feminist knowledges; or to dispense with them altogether, reverting to an anti-theoretical, anti-intellectual reliance on 'experience' or 'intuition', it remains unresolved relative to this crisis. In other words, where feminism remains committed to the project of knowing women, of making women objects of knowledge, without in turn *submitting the position of knower or subject of knowledge to a reorganization*, it

remains as problematic as the knowledges it attempts to supplement or replace.[3]

The second type of feminist theory, then, aims to recognise patriarchal investments and move towards understanding and challenging them *as the basis of knowledge*. It is 'concerned with articulating knowledges that take woman as the *subject* of knowledges.'[4] Rather than merely attempting a critique of normativity, 'these feminists have had to develop altogether different forms and methods of knowing and positions of epistemological enunciation, which are marked as *sexually* different from male paradigms.'[5] It is this second type of feminist theory that I have relied most heavily on throughout this book and the area of enquiry I have attempted to further with this project. Through my study of mundane daily practices and embodied identity, I have worked to bring together ideas and observations from the work of Elizabeth Grosz, Donna Haraway, Luce Irigaray, Karen Barad, Gilles Deleuze, and others who take patriarchal knowledge as their starting point and object of analysis. Taken together these knowledges contribute to an embodied ethics; an approach to experience, knowledge, and understanding that takes matter seriously.

CHAOS, TERRITORY, POSSIBILITY

If patriarchal reason seeks to make sense of the world and ways of life through categorical universal truths, which bound chaos through rigid systems of representation in order to *create* patterns of sameness, a new ethics of being seeks to get closer to the chaos (i.e., the material, the sensory, the becoming) by learning to recognise patterns of difference. It is derived from and through embodiment. Put simply, mainstream patriarchal knowledges work reflectively, embracing sameness and demonising or systematically subduing difference, while an embodied ethics works diffractively, becoming aware of several levels of being and acknowledging, giving value to differential ways of living.

When considering those concepts of identity explored throughout this study, reflective understandings of the self are vital for the ability of fear, anxiety, shame, and embarrassment (FASE) to work. This is because reflection for both *homo clausus* and *homines aperti* ways of being are processes of territorialization, albeit in conceptually opposite directions. Where *homo clausus* begins without territory and must perpetually produce the borders of one's body, one's territory, from the self within, *homines aperti*, instead, create their self from a constantly deterritorialized or fragmented position, creating territory through visual affirmation with other fragmented subject-objects in an ongoing, anxious attempt to 'look the part'. In both cases one is either too territorialized or too fragmented to readily experience the effects of chaos, of becoming-other, and is left stagnant, fragile, and disembodied. (I will return to FASE below.) When one who understands experience according to these identity politics does go through a threshold experience with conscious awareness—that is when the experience of difference is recognised and integrated into one's identity—it can be a period of crisis or epiphany, as well as a major life event. It can be a substantial crack in the 'structure' of one's being.

Alternatively, when being and ways of life are understood as less rigid and more pliable than those organised through *homo clausus* and *homines aperti*, such threshold experiences are not only less devastating, they are more common and thus highly generative of possibility. Diffraction for *corpus infinitum* seeks to expose 'how boundaries do not sit still' and thus are open to becoming-other.[6] Rather than creating territory onto or out of the body, *corpus infinitum* takes the materiality of the body as ontologically constant and not in need of further territorialization. Thus *corpus infinitum* recognises the differences inherent to bodies and bodily being and is open to experiences of becoming, of sensation, of change, of difference, because it is an embodiment which trusts and values the materiality of the self in and of itself. Rather than fearful, anxious, shameful, or embarrassed by materiality, *corpus infinitum* allows for the possibilities of bodily comfort, trust, creativity, and curiosity, not as something to obtain but as ontologically primary and personally vital. For *corpus infinitum*, materiality is cohesive in its pliability and thus energy does not need to be spent on the process of territorialization. When we begin to be less restricted by heteronormative sex-gender-sexuality, we can begin to trust the materiality of the body as the ontological given, and not as the passive thing which we must actively shape into an image of the self.

Throughout this book then, I have attempted to show that *corpus infinitum* is a possibility for being and that that possibility (for becoming-other) is available to us in our lives already, even (perhaps *especially*) in our most mundane daily activities. Specifically, I have suggested that public toilet spaces offer insights into how we construct embodiment, that is, the fundamental way we are in and of the world, and can therefore help us access the workings of power usually ignored in daily life. It is of no small consequence that since public toilets are sex-segregated, self-experiences within them are also specifically sex-segregated and thus give us an opportunity to disentangle the workings of power in a space common to, yet glossed over in, daily life. I have suggested that these spaces help (re)produce an unethical binary sex-gender along a singular axis of heteronormativity, therefore informing our materiality in meaning and experience, as well as carefully tracing how that occurs. With this new knowledge and awareness we can begin to recognise the need to re-conceive of how we construct, understand, and experience identity and embodiment in our (re)configuring of the world.

Before moving on to the implications of this possibility, I will briefly summarise the book both as the parts of a whole and as a whole, which—to return to an idea introduced in the theoretical chapters—can be understood as different from, more than, merely the sum of parts.

PARTS OF THE WHOLE

In chapters 1, 2, and 3 I introduced the specific theoretical approaches to identity that entangle this book. They were *homo clausus, homines aperti*, and *corpus infinitum*. While *homo clausus* and *homines aperti* are dialectically opposed, my approach, *corpus infinitum*, is non-dialectical. The *homo clausus/homines aperti* (both terms from Norbert Elias) dialectic can most simply be described as a closed/open relationship. I will summarise these in turn. The *homo clausus* is understood as the monadic, closed individual—seemingly neutral, but as we know from those feminists described above, there are no *neutral* knowledges. The ontological basis for the ideal *homo clausus* is male, and females are understood as inherently lesser and opposite to them. The *homo clausus* is highly untrusting of materiality and thus must continually territorialize itself through material processes to conceptually bound itself in order to give off an impression of stability and sameness. This process is understood to be directed from the 'self within the case'—the 'inner truth' of the bodily object. The *homo clausus* is understood as primarily a solitary being.

Alternatively, the *homines aperti* approach recognises the interrelated nature of individuals, that people are *social* beings. *Homines aperti* persons are also understood to come from a 'neutral' ontology—i.e., male—and the self is consolidated through social processes. Rather than stable, this postmodern subject is understood as fragmented, using bits and pieces of those subject-objects outside of it to create the self. That is, the process of selfhood is an ongoing process of re-territorializing the fragmented body-self. In some approaches to this model of the self, the body itself is thought to be materialised in this process. While this represents a return to materiality, it remains highly problematic because it is unclear how the process of territorialization (via discourse) creates the body. Thus the body remains 'neutral' (i.e., male) and unable to actually engage with the discursive process of territorialization. *Homines aperti* are anxiously caught in an ongoing cycle of de-territorialization; never quite stable enough to recognise the opportunities for becoming-other.

Lastly, I introduce *corpus infinitum* as a way into understanding how *homo clausus* and *homines aperti* work according to heteronormative structures of disembodied identity. Where both *homo* and *homines* mean man or men—that is, *homo clausus and homines aperti* literally and conceptually *begin* with identity—*corpus infinitum* begins with the body. In chapter 3 I posit an approach to identity that implies a fundamentally differential, non-patriarchal approach to materiality, that is, as *living*. Rather than something to be managed and controlled *in order to* enable thinking, knowing, and understanding, this approach posits that thinking, knowing, and understanding *are material processes* and that materiality is *always already active*. Thus, rather than a return to essentialism, which values categorical sameness, this approach to materiality espouses differential becomings. That is, materiality *as the source* of experience. You'll notice sameness and experience here are non-dialectical; they are markedly different approaches to the body.

After these three chapters, in chapter 4 I give a brief history of public toilet spaces. I trace how they emerged and how the corresponding bodily dispositions feed directly into *homo clausus* and *homines aperti* ways of being we continually replicate today. From this chapter we learn that both the toilet spaces themselves and our ways of feeling about our bodies while using them has changed very little since their development.

Chapter 5 is the first empirical chapter, where I introduce my study more specifically and begin to disentangle my data. This chapter unfolds according to the rules of the *homo clausus* triadic intra-action order (TIO), which I use to elucidate my data. In this chapter, I show how the TIO creates gendered experiences in equal yet opposite ways. That is to say, the rules of the TIO that govern people's behaviour are the same for both sex-genders, but are carried out in slightly different, seemingly 'opposite' ways. This is a point I return to below. This chapter illustrates how the body is construed as abject and threatening to *homo clausus* identity and thus something to rigidly manage at all times. This chapter can be understood as the most heteronormative example of public toilet use in the book.

Chapter 6 is concerned with public toilet use that acknowledges the interconnected nature of social life, which can be described as *homines aperti*. This chapter looks at practices of care that undermine or blatantly reject one or more rules of the heteronormative triadic intra-action order, but also how fear, anxiety, shame, and embarrassment are experienced individually, yet for social purposes. Vitally, though, in these examples of interdependency, *homines aperti* expose fissures in the identity structure of the *homo clausus.*

In the final empirical chapter, chapter 7, I explore the potentialities for being bodily when differential being is recognised and valued. This chapter takes practices of play, pleasure, and possibility in relation to bodily being in public toilets. These practices are direct challenges not only to the rules of the TIO, but also to the emotions we are inculcated into experiencing that help maintain the TIO and *homo clausus* ways of being generally. These differential ways of life are described as *corpus infinitum* and point to the thresholds available to us when we stop allowing the de-/re-territorializing nature of fear, anxiety, shame, and embarrassment to control our ways of being. Ultimately, this chapter points to the opportunities for becoming-other that are available to us in daily life but are overlooked or ignored because of our social systems of habit which coalesce into structures of being. This chapter champions the materiality of the body as the source of all experience, knowledge, and understanding.

THE WHOLE

As a whole, this book is highly structured, generally following a rule of three—three primary frameworks (theory), which underpin three primary examples (chapters), which unfold according to three guiding explorations (sections within chapters). To push this observation a bit further, the overarching case study (public toilets) is a singular space in two seemingly opposite manifestations. When considered from this vantage point the book is *reflective* of those representational knowledges I critique (e.g., sex/gender as singular and opposite but not ontologically different, judgemental rationality and categorisation, linear heteronormative progression). Despite this, I have worked in three important ways to entangle this territory with patterns of diffraction. As already explained, patterns of difference are not always easily recognised; rather, we have to *learn* to apprehend them. Thus I highlight each diffraction pattern entangled in this book in turn.

The first is with the concept of *corpus infinitum* itself—that is, by posing a materially directed way to approaching identity and everyday life that takes the active body, not the responsive social *self*, as primary. This concept aims to destabilise patriarchal subject positions by exposing their inherently open, pliable nature. Rather than a replacement for *homo clausus* or *homines aperti, corpus infinitum* points to the possibility of a new subject position which is not dialectical or conceptual but rather experiential and material. *Corpus infinitum* is an intervening in and not an extension of the heteronormative.

The second is the almost counterintuitive undercurrent that the empirical chapters follow. In chapter 5 I introduce the TIO as the heteronormative system which people are compelled to follow while using public toilets. This system is normative and orderly and allows for predictable use of public toilets. It also relies on socially instilled bodily fear, anxiety, shame, and embarrassment to work. It reinforces the feeling that we should not trust or be comfortable in our bodies, each time we engage with it. Thus it is the condition and the cause of the two chapters following chapter 5. Yet those chapters, chapters 6 and 7, work by explicitly undermining the power of both the TIO (the rational actions) and the emotions it imposes. This is a process of de-territorialization, of an opening to chaos from within a structure by showing how that 'structure' is easily weakened through differential ways of being.

The third is the use of poetry and narrative throughout the book. This is perhaps the most obvious intervention of difference but an important one. It is there to help expose how knowledge practices *are* material practices, that reading *is an embodied activity*. I push this further in the epilogue, which directly follows this conclusion. There, the text is designed to draw out the material entanglements of knowledge and identity and highlight the experiences of being and becoming-other.

These differences in this book should not be overlooked as mere anomaly, transgression, or unconsidered variance—that is merely the opposite of a traditional self-same approach. Rather these are interventions into ways of being and knowing in their materiality. If I had more time and space, it is possible that I could compile something which is much more diffractive, like Karen Barad's brilliant book *Meeting the Universe Halfway: Quantum Physics and the Entanglement of Matter and Meaning*, where she takes a thoroughly diffractive approach to the text itself. Using that text as a point of departure, in the next and final section, I will outline the implications of this study and a few prospects for future entanglements.

ENTANGLED FUTURES

Corpus infinitum, rather than a new identity structure, points to ways of becoming more thoroughly embodied which are already available to us in daily life across all identities. By taking the materiality of the body as ontologically primary—the location of all social, mental, emotional, philosophical processes—our ways of life can become more comfortable, creative, curious, and trusting, as well as less riddled with destabilising fear, anxiety, shame, and embarrassment. I view this kernel of difference—this one case study—as part of a much larger and more densely entangled possibility. While I cannot elaborate at length on this possibility here, I will at least sketch an outline of what I have not been able to include in this text and suggest a few additionally pertinent outlets for consideration.

I consider *corpus infinitum* to be a particular narrowing of Karen Barad's (2008, 2007, 2003) more general approach to the human that she terms agential realism. This approach understands personal agency not as something that a subject *has* or holds within the body, but rather part of a relational process that allows opportunities to emerge. This is because the body is neither passive nor fixed, not waiting for culture to sculpt it nor a thing of nature prior to culture. Instead, 'matter is always already an ongoing historicity.'[7] When matter is understood in this way it means we must also adjust our ways of understanding and knowledge making to account for this ongoing historicity. For example, when developing theory (Barad's example is quantum theory),

> rather than giving humans privileged status in the theory, agential realism calls on the theory to account for the intra-active emergence of 'humans' as a specifically differentiated phenomena, that is, as specific configurations of the differential becoming of the world, among other physical systems. Intra-actions are not the result of human interventions; rather, 'humans' themselves emerge through specific intra-actions.[8]

This is basically what I have attempted to show throughout this study with my specific focus on the sex-gendered body and identity—that is, by entangling theory and practice and by practicing the theories of those scholars I admire and draw from. This approach to understanding matter and the 'human' can be described as posthumanist. Posthumanism recognises that 'we humans' are not outside of the world, we are not a supplement to nature but a natural phenomenon, just like other natural phenomena. When we begin to reconceptualise the human, we can begin to reconceptualise human ways of life. One way of life worth rethinking is patriarchal capitalism.

In February 2012 I heard Elizabeth Grosz speak (at the American Association of Geographers annual meeting in New York City) on and around her text *Chaos, Territory, Art: Deleuze and the Framing of the*

Earth.[9] Something she spoke about remains with me, which is how our emotions and sensations are pre-digested for us by capitalism. This is an idea present in my study both specifically and generally. Specifically with the four conditioned emotional responses (FASE) I focus on, and generally by connecting those emotions to our social ways of being— that is, how they are intra-acted. As individualism is necessary for capitalism and *homo clausus* ways of being, which are still the ontological foundation to our identity construction, are specifically individual, I view this project as a way to critique and problematise capitalist ways of being from an embodied perspective. That is to say, in many ways I believe *homo clausus* is the first inculcation not only into capitalist subjectivity (i.e., individualism) but also capitalist labour. In other words, *homo clausus* is our first lesson in capitalist labour insofar as it is the condition of individual subjectivity.

The French poet, Arthur Rimbaud (1854–1891), wrote in the 'Mauvais Sang' section of his prose work *Une saison en enfer* (A Season in Hell) 'I have a horror of all trades.'[10] Kristin Ross, in her excellent book *The Emergence of Social Space: Rimbaud and the Paris Commune*, relates this anti-work expression directly to subject formation. She explains, 'The regime of work, then, is inseparable from the development of form, to which corresponds the formation of the subject.'[11] Much of Rimbaud's work is concerned with this anti-labour theme and particularly the overtaking and dividing of individuals through the morality of work.[12] One of his famous quotes, also from *Une saison en enfer*, apropos to this discussion is 'I'm intact, and I don't care.'[13] This is related to the idea expressed in chapter 1 with sensorial individuation. That is to say, when we de-territorialize our body according to capitalist ways of being we necessarily divide ourselves and are no longer 'intact' as whole, cohesive beings, but are instead separated into parts with particular functions. Ross explains this through Rimbaud's work. She says to have a trade 'is to lose one's hand as an integral part of one's body: to experience it as extraneous, detachable, in service to the rest of the body as synecdoche for the social body, executing the wishes of another.'[14] In the contemporary West, a life of labour is expected for most (who are not the super wealthy) and it is my suggestion that it is through our identity construction, persistent individualism—taken as a given, but which I have worked to show is an ongoing process of naturalisation—that we are primed for participation in and support of patriarchal capitalism. Thus, if we want to critique patriarchy and capitalism in fertile ways, we must begin by understanding identity, and in order to understand identity we must be able to approach matter onto-epistemologically as active and relational. There are a few fruitful ways into such a critique and into the matter of everyday life.

This approach to matter can be applied diffractively along with other methods and knowledges in order to draw out the emergent patterns of differential becoming coalescing in daily life. There are three areas of exploration that could benefit most readily from such an approach. Those are: science, technology, and related processes of globalisation; body and place in the post-colonial experience; and the development and application of social work (including care) practices and the drafting of social policy. These venues could mean a greater entanglement of disciplines where the humanities and sciences could more readily speak to and influence one another. This is in line with Grosz's work where she elucidates how art and science share the same (vibratory) force of chaos yet approach it differently.[15] Where science creates rational, predictable patterns, art creates sensation. Once we understand and respect this fundamental, ontological point of departure it may be easier to entwine those disciplines that have been traditionally considered disparate. Thus it may become possible to cultivate a pliable, ever-ready embodied ethics of being.

NOTES

1. Elizabeth Grosz, *Space, Time and Perversion: Essays on the Politics of Bodies* (New York: Routledge, 1995), 39.

2. Grosz, *Space, Time and Perversion*, 39.

3. Grosz, *Space, Time and Perversion*, 40.

4. Grosz, *Space, Time and Perversion*, 39.

5. Grosz, *Space, Time and Perversion*, 41.

6. Karen Barad, *Meeting the Universe Halfway: Quantum Physics and the Entanglement of Matter and Meaning* (Durham, NC: Duke University Press Books, 2007).

7. Barad, *Meeting the Universe*, 151.

8. Barad, *Meeting the Universe*, 352.

9. An audio recording of the talk and discussions is available here: http://societyandspace.com/2012/04/19/elizabeth-grosz-discussion-at-the-aag-audio-recording/.

10. Kristin Ross, *The Emergence of Social Space: Rimbaud and the Paris Commune* (London: Verso, 2008), 50.

11. Ross, *The Emergence of Social Space*, 50.

12. Some have speculated that this is the reason that he stopped writing altogether at such a young age of just twenty years.

13. Ross, *The Emergence of Social Space*, 50.

14. Ross, *The Emergence of Social Space*, 51.

15. Elizabeth Grosz, *Chaos, Territory, Art: Deleuze and the Framing of the Earth* (New York: Columbia University Press, 2008).

Epilogue

"AND IN A SENSE, THAT IS MY JOURNEY TO SHIT TOO."

I'm a gay man, I'm male; I know this now. I can say am I this or that, but it is different for me every day. My body is different every day and I'm beginning to rethink things a little, I'm beginning to question some of those labels. I definitely recognise that there should be more fluidity in how we define and experience identity, but I am unsure of how to participate in an active redefinition, redefining, other than just living the way I want to live and trying not to be constrained.

So, in the spectrum of masculine to feminine, I'm more comfortable with straight men who are comfortable straight men, who are non-aggressive, who are not clinging to an identity based on hypermasculine, heteronormative sexuality; they are really my favourite people in the world. I love male energy and I love when it is friendly and open, and hostile male energy is still the scariest thing to me. Because I had so many incidents as a kid and I was such a feminised child and I was so anti-male growing up and so expectant of abuse that sometimes came and sometimes didn't, that it is still something I feel very tenuous about.

So, I guess my desire for fluidity is a journey toward acceptance of myself and others, and that is absolutely centred on my body. For instance, now, I have a very open relationship to poop, and that is a purposeful thing for me.

I remember when I started acting classes and they said we are going to think about our bodies and look at and feel them and we are not going to judge; we are just going to notice. *That was revolutionary for me!* Now, in retrospect, I can think about this, about my life in terms of space. I can locate how I felt and try to understand how I became who I am through specific places and how that continues to shape me every day. I have a lot of history of safe spaces versus not safe spaces, and the space of the toilet fluctuates between the two. And what I'm understanding now is that my relationship to public toilets, and toileting in general, is part of a much larger story of me becoming me.

When I was younger I definitely wanted to be a woman and definitely fantasised about that. I didn't think it was practical, it was more what I saw myself as already, what I wanted to be; when I grew up I wanted to be this beautiful woman and I'm pretty sure I knew that wasn't really going to happen, but it didn't stop me from living my life in such a way that even as a child people recognised that desire. Twice when I was on the playground at school, I remember kids who were not in my normal social circle, who were not at my school, kids at a strange playground, coming up to me and asking me if I was a boy or a girl, because I was really feminine; I had really taken that on. I remember, my friend Jason, he was my best friend growing up and we were in this performing group together, and it is so funny because now that I'm finally thinking about bathrooms as these spaces, these safe spaces again, it makes me think of him.

See, Jason and I would hang out in the bathroom; it was where we would kind of powwow, and of course later we would both be gay men, so the bathroom was where we bonded. I remember being at this festival, our group was performing there, and we were hanging out in the bathroom, the two of us sitting there judging people and their cleanliness habits through this hand-washing hierarchy: from soap and hot water—that was the best, and then with cold water and soap, and then warm water no soap, then cold water, and lastly no hand washing, which we thought was gross. . . . So washing your hands was important, and if you didn't, it was naughty, naughty, naughty. . . .

And a lot of what grossed me out and fuelled this desire to be a woman was that I was really disgusted by these things that represented masculinity, in the way that I thought it was gross that guys burped and how they peed in the toilet. I didn't want to get hair on my legs, I thought that was horrifying. The gender issues and identification issues were wound up with bathroom spaces; they were certainly connected to bathroom spaces; they were *automatised* into bathroom spaces. So men's public toilets were already these spaces imbued with disgust for me as a kid, but they were also important spaces where my friend and I could distance ourselves from that by watching and judging other people, by holding them to our standards of cleanliness, and that helped me feel okay about not being a girl.

When I was younger I had encopresis, which is a term I just learned a couple months ago, and I'm not sure at what age it started—I'm trying to remember it for myself at this point as well for research material for a show I'm writing about my life and poop and just as an unearthing, as my own 'who the fuck are you?' process of life. It is hard for me to track the history because I remember only blips, moments, so I'm forced to fill in the rest or continue to mine for data somehow. My parents don't seem to remember ANY of this. I think they blocked it out, but they also had blinders on about me being queer for a long time—I had to make it explicit, it had to be made explicit—ironically because they found pornography on my computer and it was like the tables have turned—*wow, wow*—I just made a fascinating link—what is really interesting about that story (which is separate from this) is that, I was really into the pornography and I was hiding it and they would find it and other things that would happen. But it wasn't until it was explicitly gay and couldn't belong to anyone else that they put a stop to it and made it a big deal. There were other times when I was growing up that seemed like they were like, 'we saw this thing, we know that it is there,' but they wouldn't punish me until it was definitely gay and so there was this whole guilt cycle about them finding something but not punishing me for it. . . .

So, encopresis was something that I dealt with through high school. Like I have this one incident that was the pinnacle that happened when I was in Spain the summer after my freshman year of high school. Where I didn't go, I didn't poop, for two weeks. Two weeks. It was to the point where I couldn't eat anymore. I couldn't take anything in; I was just full. I remember my host mother would make these omelettes and things for dinner and I would have to wait for her to leave the room and I would put it in a bag, because I just couldn't do it, I couldn't eat and it was so shameful. But the process is very much about control, and for me, again, there is this whole experience of safe spaces and not safe spaces, which for lack of, well, what my understanding of encopresis is about a fear of the feeling, the sensation of going poop, which may have factored in, but that doesn't seem to have been the main thing for me; it was more fear of the toilet itself as dirty and the flushing and the letting go and what I had inside me as dirty, and these feelings connected to my masculinity and my sexuality too.

What I seem to remember, those markers that stand out to me, the only way that I am able to trace this history is through shame, like the times when I clogged the toilet as a kid and was made to feel really badly about that. It was gross, it was an inconvenience and as a kid I couldn't deal with it, I couldn't fix it so of course it was outsourced to my father, who was not a gentle—I suppose if we want to make the link, I'll do the work for you of analysing me, which isn't your job, it is for my analyst!—if we want to make the link between my gender and those issues and the toilet, my father was also the main person who gave me

negative feedback about homosexual desires, and throughout my childhood, I definitely preferred my mother, I was a mama's boy and felt
alienation from my father that I definitely felt more of as I grew older. I
didn't let my dad teach me how to shave because I didn't want to have
to start shaving; I was horrified that it was going to happen at all.

So, that shame about my body, my desires, my needs was very
strongly felt, and I didn't go in Spain because I didn't want to clog the
toilet I guess or because it was a strange environment, I don't know why
exactly, but I'm sure fear and shame were a part of it.

So, what would happen and kind of the cycle for me was definitely
about safe space. I learned at some point that when I had to go, I had to
go, but there was also pleasure in the controlling of not going, *wow*, and
in controlling it and in the actual feeling of it. I think it was both
physically pleasurable and there was pleasure, satisfaction in the control
itself and I think I also knew that it probably wasn't good. That I was
doing something that was not right, well, that isn't the word I want to
use . . . that it was unhealthy. Like unhealthy in a 'you're different' way,
like you're doing something other people don't do, and unhealthy in a
'this is bad for your body' way. This is all still fresh for me. . . .

So I wouldn't shit at school, and there was a hierarchy in my own
house as to where I would go, ya know, I don't know, I must've gone
back and forth on that. . . .

I remember another time when Jason came over to our house and
he had the same thing, no, not the same thing, but a nervousness about
public spaces, I don't know if he withheld like me, but that he wanted to
clean the toilet before he sat down to go in our house. Which I was kind
of like, "but it is my house Jason it's clean!" A little upset by that. And I
remember going at his house once, and my grandparents' houses were
okay, well, my maternal grandparents' house was okay. But I wouldn't
go anywhere else at all.

The thing with my grandparents was that I felt very close to them,
especially my maternal grandmother, who I suppose was probably the
most positive influence on my homosexual tendencies; she is the person
who pulled my mother aside on her deathbed and said, your son is gay,
you need to know—aw, aw, grandma—so I would go in her bathroom.
They had this bathroom which had big mirrors, a shower with a sliding
door which is frosted, the toilet in the corner and the sink with all sorts
of perfumery things on there—this was the show I was writing before I
decided to do the poop show. A couple years ago I was writing a show
about my childhood desires to be a woman and I saw my first pornographic films when I was seven or eight, which kind of factored into my
modelling of sexual behaviours—I mean I am like a big ball of complicated but exciting issues for people like you and for people like me

too—I'm fascinated by it, like how the fuck did I come out like this? So, I remember that that was the place where I would empty myself and the mirror maybe helped, but I would stick my fingers in my butt and get poop out and look at it and see how it worked and sitting there straining for something to come out because otherwise I would be holding.

Maybe because my grandparents just never asked what I had been doing in the bathroom for the last twenty-five minutes, I don't know, but that is where I felt comfortable, and otherwise I would try to hold it. And I would hold it in school, and in middle school I did this production of *Joseph and the Amazing Technicolor Dreamcoat* that came through town, and I was part of the chorus. And I remember messing up my costume pants, my shorts, and trying to clean it myself and being so embarrassed. And I remember this one time, this is so awful, when I was in this dance studio again with the performing group and now that I think about it, I remember I kept a spare pair of underwear with me, or I would put toilet paper in my shorts, or between my cheeks. I remember I would smell and other people would be like, 'Ooh what is that?' And being like oh my god it is me, I know it is me! So, I would take my underwear off and stash it in my bag or something or wait until I got home and cleaned it. Oh, oh the story! So I was at this rehearsal and a little doodle fell from my pant leg onto the floor and I was like, OH MY GOD, NO! and I smushed it into the ground with my shoe, in this dance studio. I was in middle school, but young because I skipped a grade and that is what is so fascinating to me about it, about my history, because I excelled and was brilliant in so many other ways as a child and really supported and it is so funny that these underbelly issues which I understood I had, but didn't really know how to deal with and it wasn't like this public family issue, at least they claim to not remember *any* of this even though I struggled with it for a long time.

The way I remember my family being involved, well, we went to this doctor and I got put on a laxative and I tried it once and it made me gag and so we didn't do it, I didn't take it, because I was a prized child, I was golden—*what baby doesn't want, baby doesn't get*—I was pretty snotty actually . . . and then I went on a high-fibre diet because they thought it was about regularity, and the kid I know—I recently taught a class of toddlers and had a kid who was dealing with it and I don't remember dealing with it at the age of two, his seems to be psychologically associated, but also not. His little brother was just born and it seemed to start at the same time, and he would wail and scream and come into the class and clutch his bottom and thrust his hips forward and be screaming but no, 'I don't want to go to the potty'. I don't remember ever being like that and my parents don't either. They don't remember the laxatives, the high-fibre diet, and they don't seem to remember finding the poop—the dirty underwear, which I either cleaned out myself or threw away or hid for a while. And there is an interesting link there for me about fluids, because I remember seeing the pornography and masturbating at a really early age, like seven or eight, and so as soon as my body was capable of producing semen I was ejaculating onto things which would then go into the wash or I would wash myself and that was never discussed, kind of like the porn that wasn't gay. My mother did the laundry and she was very thorough, she would've known. She knew. I had this silk robe and the whole point of this silk robe, I mean A., I loved it because I felt feminine and beautiful in it—it was green with a black stripe, like jade—but also, B., because I would have sex with it at night and blow my wad into it. I would take it into the bath with me and wash it.

So the place where I would let the poop out was the shower, and my parents would be involved in the times when I would clog the toilet and my father would have to plunge it when I was home or, when we were in California for two weeks for pilot season as a child actor, my mother and I were living in this apartment and I took one that was so big I clogged the toilet and we couldn't use it for like three days, so my mom was like shitting in jars, so it did become a family issue in that sense. She remembers that. I also remember one time we were at a hotel, or a motel, and my father was like, 'you have to go get the plunger, *you did this*', which has this whole association, contamination thing—like if I'm the one asking for the plunger, I must've been the one who did it.

So, that seems to me to be the feedback cycle, which I think, again, is related to this whole issue of the person I wanted to be, who is above all of that, who is feminine and beautiful and who is not dealing with shit. And I remember at some point realising that I spent so long holding it in and that I couldn't hold it in any longer, that it had to come out and I had the urge, and I knew I would get constipated if I had the urge but wasn't in a place where I could go. Oh my god, maybe that is why I have hip tension—I have these spots that are really stuck and I wonder if it has to do with all of those years of holding it in, hmmm. . . . So, at my desk at school and I couldn't go there, so I had to wait until I got home, and once I was home, I needed to get the urge. So, I had to learn that when I had the urge I had to overtake it. And sometimes I would get the urge in the shower and let it come out and that is where I remember not allowing myself to go fully, I would just go a little bit because I knew the shower wasn't the right place, but I knew it was an in-between place. I would go and smash it into the drain and it would smell more than when I went in the toilet of course.

And finally, in high school I started to be able to poop in public toilets. I remember one experience in California, before high school, which would have probably been sixth grade, where I had diarrhoea really badly, and I had to go in a public toilet and that was like an intensely charged experience because I *had never done it before*, that is why I remember it so starkly and then, after that one time, not really doing it again until high school. I remember moments in high school where I became aware of other people shitting and going into a stall and seeing poop. So, one of the most important discoveries that happened in high school was hearing a friend of mine, Laura, saying "oh I have to shit" or "oh I just shat and it is really stinky," and me being like, oh my god! We're in someone else's house and she just pooped here!

I don't think I had a problem with peeing, although I don't like trough urinals, I get pee shy, I know even now if there is a possibility of someone being next to me as I pee I won't be able to go. And in crowded male situations like football games, trough urinals horrify me, they always have. And public showers, which I have gotten better with, but it is something I breathe and talk myself through, and I know it is all about bodily shame. So as I get more comfortable with my body it all becomes easier, and the greater issue I'm dealing with is validation of my body as an okay public body, as a sexy body, as acceptance from and for my body, and just not to hide. I have been struggling with this acceptance for a long time, even as a kid. I remember my father saying, 'what you think you're so special? You think what you've got is so special?' and just being like, no, that is not it—I didn't think I or my body was better than anyone else's; he didn't understand why I was so closed. And wrestling with this idea of why people wear clothes, we are all naked, and just not being able to grasp all of these things. And with that, those social mores about shitting, I just didn't know how to navigate them because I saw no public context that mirrored them, that said it is okay to go, or this is safe, or other people do it. I remember wanting, but feeling grossed out, but wanting to find poop in public toilets, wanting to be like, oh, there it is, *there is the evidence*. I felt so alone in this. And I think by that time, by high school, I had started doing it because Spain was a transformative event for me. It wasn't instantaneous, but I knew I couldn't do this anymore. I had to be more open and accepting of myself. I had to allow myself to be empty and to stop holding onto an idea of who I was.

So, in Spain, I waited two weeks and finally went after I slipped it out in the shower and being totally uncomfortable for two weeks and the difference was that my host mother, Maria Teresa, was so generous about it and with me and so warm and was like, 'oh my god this is why you've felt so awful' and 'here we're just going to pour water on it until it breaks up' and 'it is okay, you have to go to the bathroom' and all of these things I never knew or believed about myself or could be possible and I was like, 'OH, Okay, thank you, I will try.' She wasn't angry, or placing any blame on me, or humiliating me. She just wanted me to be okay. To feel safe.

So in high school I was older and more aware and able to deal with it, I think, because I was in the drama program and we had these dressing rooms with bathrooms in them and I found my places where I could go and be comfortable but also, I remember being conscious of people not washing their hands and being grossed out and these things and feelings I explored early on with Jason as being present. And that was right around when I was figuring out my sexuality. Earlier in high school I was like, 'maybe I'm bisexual, maybe I don't have to be gay, maybe I like girls enough and I can just ride on that', and then I was like 'Nooo' . . . so there were a lot of other things I was figuring out and I was ready to admit to other people that I was going to call myself gay, and that was like sophomore year of high school.

[*Does when you were able to start pooping in public coincide with when you were able to start saying, 'I'm gay'?*]

Yes, I suppose it does. I suppose it does. WOW. *Look at that.*

I think it is important to realise that all of these things about identity, about being, are fluid and flowing. And in the spectrum of masculine to feminine there are still places where they are very muddy for me and I still hierarchise, like at work, but my personal comfort came to really accepting, or is a *process* of accepting who I am, how I feel, and letting that exist, and in a sense that is my journey to shit too. That is, that it isn't really about anything, it just is, it is a collection of particles but of course there are codifications. And I see my own constructed persona as very heteronormative to some extent and that is intended, I want to be able to float between both. But in the world of constructed gender with masculine dirt and feminine cleanliness, where does that leave us? And sometimes when I consider going into women's bathrooms at bars, I go through this process of telling myself, 'You're not dirtier than women. This barrier is artificial', and for the elements of it that are sanitary versus gender comfort, I remind myself that I do not abide! I am just as clean! I say this to myself. It helps. And now, often I wonder if I shouldn't be more concerned about certain things, like I'll pee or poop in a toilet that already has urine in it because I don't want to waste the water to flush it, or I'll just wipe off wetness on the seat, or I sit down and realize it is a little wet, and I just don't really care. I've become really relaxed about it. And that is the thing with my progression with pooping in public, I will not put my body through what I used to do and I want to change the way that people talk about it and the presence it has in society to one not where we have to talk about it all the time but where poop-positive space exists, where we don't have to wrinkle our noses every time. I just think it is important to realise that all of these things are fluid, flowing, and ongoing.

BIBLIOGRAPHY

Ahmed, Sara. *Queer Phenomenology: Orientations, Objects, Others*. Durham, NC: Duke University Press, 2006.

Alaimo, Stacy, and Susan Hekman. *Material Feminisms*. Bloomington: Indiana University Press, 2008.

Alcoff, Linda. 'The Problem of Speaking for Others'. In *Who Can Speak: Authority and Critical Identity*, ed. Judith Roof and Robyn Wiegman. Urbana: University of Illinois Press.

Back, Les. *The Art of Listening*. New York: Berg Publishing, 2007.

Barad, Karen. 'Quantum Entanglements and Hauntological Relations of Inheritance: Dis/Continuities, SpaceTime Enfoldings, and Justice-to-Come'. *Derrida Today*, 3, no. 2 (2010).

———. 'Posthumanist Performativity'. In *Material Feminisms*, ed. Stacey Alaimo and Susan Hekman. Bloomington: Indiana University Press, 2008.

———. *Meeting the Universe Halfway: Quantum Physics and the Entanglement of Matter and Meaning*. Durham, NC: Duke University Press Books, 2007.

———. 'Posthumanist Performativity: Toward an Understanding of How Matter Comes to Matter'. *Signs*, 28, no. 3 (2003).

Bartky, Sandra. 'Foucault, Femininity, and the Modernization of Patriarchal Power'. In *Writing on the Body: Female Embodiment and Feminist Theory*, ed. Katie Conboy, Nadia Medina, and Sarah Stanbury. New York: Columbia University Press, 1997.

Bataille, Georges. *Eroticism: Death and Sensuality*. San Francisco: City Lights Books, 1986.

Bordo, Susan. *Unbearable Weight: Feminism, Western Culture, and the Body*. Berkeley: University of California Press, 2003.

———. *The Male Body: A New Look at Men in Public and in Private*. New York: Farrar, Straus and Giroux, 2000.

———. 'The Body and the Reproduction of Femininity'. In *Writing on the Body: Female Embodiment and Feminist Theory*, ed. Katie Conboy, Nadia Medina and Sarah Stanbury. New York: Columbia University Press, 1997.

Bourdieu, Pierre. *Distinction: A Social Critique of the Judgement of Taste*. Cambridge, MA: Harvard University Press, 1984.

Bradantan, Costica. 'Scaling the 'Wall in the Head'. *New York Times Blogs, Opinionator*, November 27, 2011. http://opinionator.blogs.nytimes.com/2011/11/27/scaling-the-wall-in-the-head/.

Browne, Kath. 'Genderism and the Bathroom Problem: (Re)Materialising Sexed Sites, (Re)Creating Sexed Bodies'. *Gender, Place and Culture*, 11, no. 3 (2004).

———. 'A Right Geezer Bird (Man-Woman): The Sites and Sights of "Female" Embodiment'. *ACME: An International E-Journal for Critical Geographies*, 5, no. 2 (2006).

Bruner, Jerome S. 'Social Psychology and Perception'. *Readings in Social Psychology*, 3rd edn. (1958).

Butler, Judith. *Undoing Gender*. New York: Routledge, 2004.

———. *Gender Trouble: Feminism and the Subversion of Identity*. New York: Routledge, 1999.

———. 'Performative Acts and Gender Constitution: An Essay in Phenomenology and Feminist Theory'. In *Writing on the Body: Female Embodiment and Feminist Theory*, ed. Katie Conboy, Nadia Medina, and Sarah Stanbury. New York: Columbia University Press 1997.

———. *Bodies That Matter: On the Discursive Limits of 'Sex'*. New York: Routledge, 1993.

———. 'Performative Acts and Gender Constitution: An Essay in Phenomenology and Feminist Theory'. *Theatre Journal*, 40, no. 4 (1988).

Cavanagh, Sheila. *Queering Bathrooms: Gender, Sexuality, and the Hygienic Imagination*. Toronto: University of Toronto Press, 2010.

Cerulo, Karen A. *Culture in Mind: Toward a Sociology of Culture and Cognition*. New York: Routledge, 2002.

Colebrook, Claire. *Deleuze and the Meaning of Life*. London: Continuum, 2010.

———. 'How Queer Can You Go? Theory, Normality and Normativity'. In *Queering the Non/Human*, ed. Noreen Giffney and Myra J. Hird. Hampshire, UK: Ashgate, 2008.

Corbin, Alain. *The Foul and the Fragrant: Odour and the French Social Imagination*. Cambridge, MA: Harvard University Press, 1986.

Creed, Barbara. 'Lesbian Bodies: Tribades, Tomboys, and Tarts.' In *Sexy Bodies: The Strange Carnalities of Feminism*, ed. Elizabeth Grosz and Elspeth Probyn. New York: Routledge, 1995.

Deleuze, Gilles, and Félix Guattari. *Anti-Oedipus*. Trans. Robert Hurley, Mark Seem and Helen R. Lane. London and New York: Continuum, 1972.

Derrida, Jacques. *Rogues: Two Essays on Reason*. Stanford, CA: Stanford University Press, 2005.

Derrida, Jacques, and Anne Durfourmantelle. *Of Hospitality*. Stanford, CA: Stanford University Press, 2000.

Edelman, Lee. *No Future: Queer Theory and the Death Drive*. Durham, NC: Duke University Press Books, 2004.

———. 'Men's Room'. In *Stud: Architectures of Masculinity*, ed. Joel Sanders. New York: Princeton Architectural, 1996.

Elias, Norbert. *The Civilizing Process: Sociogenetic and Psychogenetic Investigations*. Oxford: Blackwell, 2000.

———. *The Society of Individuals*. London: Continuum, 1991.

———. *What Is Sociology?* New York: Columbia University Press, 1978.

Feinberg, Leslie. *Transgender Warriors: Making History from Joan of Arc to Dennis Rodman*. Boston: Beacon Press, 1996.

———. *Trans Liberation: Beyond Pink or Blue*. Boston: Beacon Press, 1998.

Fiske, Susan T., and Shelly E. Taylor. *Social Cognition: From Brains to Culture*. New York: McGraw-Hill Higher Education, 1991.

Foucault, Michel. *Madness and Civilization: A History of Insanity in the Age of Reason*. New York: Random House, 1988.

———. *Power/Knowledge: Selected Interviews and Other Writings, 1972–1977*. New York: Vintage, 1980.

Frank, Arthur. *The Wounded Storyteller: Body, Illness, and Ethics*. Chicago: University of Chicago Press, 1995.

Freud, Sigmund. 'Creative Writers and Day-Dreaming'. In *Criticism: The Major Statements*, ed. Charles Kaplan. New York: St. Martin's, 1991.

Friedman, Asia. 'Toward a Sociology of Perception: Sight, Sex, and Gender'. *Cultural Sociology*, 5, no. 2 (2011).

Gatens, Moira. *Imaginary Bodies: Ethics, Power and Corporeality*. New York: Routledge, 1996.

George, Rose. *The Big Necessity: The Unmentionable World of Human Waste and Why It Matters*. New York: Metropolitan Books, 2008.

Giffney, Noreen. 'Queer Apocal(o)ptic/ism: The Death Drive and the Human'. In *Queering the Non/Human*, eds. Noreen Giffney and Myra J. Hird. Hampshire, UK: Ashgate, 2008.

Goffman, Erving. *The Goffman Reader* (Vol. 7), eds. Charles Lemert and Ann Branaman. London: Blackwell, 1997.

———. *Frame Analysis: An Essay on the Organization of Experience*. Boston: Northeastern University Press, 1986.

———. 'The Interaction Order: American Sociological Association, 1982 presidential address'. *American Sociological Review*, 48, no. 1 (1983).

———. *Relations in Public: Microstudies of the Public Order*. New York: Basic Books, 1971.

———. *Stigma: Notes on the Management of Spoiled Identity*. New York: Prentice-Hall, 1963.

———. 'Embarrassment and Social Organization' *American Journal of Sociology*, 62, no. 3 (1956).

Greed, Clara. *Inclusive Urban Design: Public Toilets*. Oxford: Elsevier, Architectural Press, 2003.

————. 'Public Toilet Provision for Women in Britain: An Investigation of Discrimination against Urination'. *Women's Studies International Forum*. Elsevier, 18, no. 5 (1995).

Grosz, Elizabeth. *Becoming Undone: Darwinian Reflections on Life, Politics, and Art*. Durham, NC: Duke University Press, 2011.

————. *Chaos, Territory, Art: Deleuze and the Framing of the Earth*. New York: Columbia University Press, 2008.

————. *Space, Time, and Perversion: Essays on the Politics of Bodies*. New York: Routledge, 1995.

————. *Volatile Bodies: Toward a Corporeal Feminism*. Bloomington: Indiana University Press, 1995.

Halberstam, J. Jack. *The Queer Art of Failure*. Durham. NC: Duke University Press, 2011.

————. *Female Masculinity*. Durham, NC: Duke University Press, 1998.

Haraway, Donna. *When Species Meet*. Minneapolis: University of Minnesota Press, 2008.

————. 'The Promises of Monsters: A Regenerative Politics for Inappropriate/d Others'. In *Cultural Studies*, ed. Lawrence Grossberg et al. New York: Routledge, 1992.

————. 'Situated Knowledges: The Science Question in Feminism and the Privilege of Partial Perspective'. *Feminist Studies*, 14, no. 3 (1988).

Hekman, Susan. 'Constructing the Ballast: An Ontology for Feminism'. In *Material Feminisms*, eds. Stacey Alaimo and Susan Hekman. Bloomington: Indiana University Press, 2008.

Huizinga, Johan. *Homo Ludens: A Study of the Play Element in Culture*. Boston: Beacon Press, 2003.

Humphreys, Laud. *Tearoom Trade: A Study of Homosexual Encounters in Public Places*. London: Duckworth, 1970.

Inglis, David. *A Sociological History of Excretory Experience: Defecatory Manners and Toiletry Technologies*. Lewiston, NY: Mellen, 2001.

Irigaray, Luce. *I Love to You: Sketch of a Possible Felicity in History*. London: Routledge, 1996.

Jeyasingham, Dharman. '"Ladies and Gentlemen": Location, Gender and the Dynamics of Public Sex'. In *In a Queer Place: Sexuality and Belonging in British and European Contexts*, eds. Kate Chedgzoy, Emma Francis, and Murray Pratt. Hampshire, UK: Ashgate, 2002.

Kira, Alexander. *The Bathroom*. New York: Viking Press, 1976.

Kristeva, Julia. *Powers of Horror: An Essay on Abjection*. New York: Columbia University Press, 1982.

Lacan, Jacques. *Ecrits: A Selection*, trans. Alan Sheridan. London: Routledge, 1989.

Lambton, Lucinda. *Temples of Convenience and Chambers of Delight*. Stroud, UK: Tempus Publishing Limited, 2007.

Law, John, and John Urry. 'Enacting the Social', published by the Department of Sociology and the Centre for Science Studies, Lancaster University, Lancaster LA1 4YN, UK, at http://www.comp.lancs.ac.uk/sociology/papers/Law-Urry-Enacting-the-Social.pdf.

Leder, Drew. *The Absent Body*. Chicago: University of Chicago Press, 1990.

Lee, Jennifer 8. 'Ejection of a Woman from a Women's Room Prompts Lawsuit'. *New York Times* [online], (City Room) 9 October 2008. Available from: http://cityroom.blogs.nytimes.com/2007/10/09/ejection-of-a-woman-from-a-womens-room-prompts-lawsuit/.

Lefébvre, Henri. *The Production of Space*, trans. Donald Nicholson-Smith. Oxford: Blackwell, 1991.

Martin, Emily. *The Woman in the Body: A Cultural Analysis of Reproduction*. Boston: Beacon Press, 2001.

Maryanski, Alexandra, and Jonathan H. Turner. *The Social Cage: Human Nature and the Evolution of Society*. Stanford, CA: Stanford University Press, 1992.

Mauss, Marcel. 'Techniques of the Body'. *Economy and Society*, 2, no. 1 (1973).

Menell, Stephen. *Norbert Elias: Civilization and the Human Self-Image*. Oxford: Blackwell, 1989.

Molotch, Harvey. 'The Rest Room and Equal Opportunity'. *Sociological Forum*, 3, no. 1 (1988).

————. *Toilet: Public Restrooms and the Politics of Sharing*. New York: New York University Press, 2010.

Moore, Sarah, and Simon Breeze. 'Spaces of Male Fear: The Sexual Politics of Being Watched'. *British Journal of Criminology*. Advance access published August 9, 2012, doi:10.1093/bjc/azs033.

Morris, Michael, and Cate Sandilands. 'Eco Homo?' Unpublished performance script from Staging Sustainability, 20 April 2011, Toronto: York University.

Munt, Sally R. 'Orifices in Space: Making the Real Possible'. In *Butch/Femme: Inside Lesbian Gender*, ed. Sally R. Munt and Cherry Smyth. London: Continuum, 1998.

Neff, Walter Scott. *Work and Human Behavior*. Piscataway, NJ: Transaction, 1985.

Nisbett, Richard E., and Takahiko Masuda. 'Culture and Point of View'. *Proceedings of the National Academy of Sciences of the United States of America*, 100 (2003).

Penner, Barbara. 'Female Toilets: (Re)Designing the Unmentionable'. In *Ladies and Gents: Public Toilets and Gender*, ed. Olga Gershenson and Barbara Penner. Philadelphia: Temple University Press, 2009.

———. 'Researching Female Public Toilets: Gendered Spaces, Disciplinary Limits'. *Journal of International Women's Studies*, 6 no. 2 (2005).

Plaskow, Judith. 'Embodiment, Elimination, and the Role of Toilets in Struggles for Social Justice'. *Cross Currents*, 58, no. 1 (2008).

Phelan, Peggy. *Unmarked: The Politics of Performance*. London: Routledge, 1993.

Ross, Kristin. *The Emergence of Social Space: Rimbaud and the Paris Commune*. London: Verso, 2008.

Saunders, Karen. 'Queer Intercorporeality: Bodily Disruption of Straight Space' (a thesis submitted for the degree of Master of Arts in Gender Studies at the University of Canterbury, Christchurch, Aotearoa/New Zealand, 2008.

Scheff, Thomas J. 'Shame and Conformity: The Deference-Emotion System'. *American Sociological Review*, 53, no. 3 (1988).

Shilling, Chris. *The Body and Social Theory*. 3rd edn. London: Sage Publications, 2012.

———. *The Body in Culture, Technology and Society*. London: Sage Publications, 2005.

———. 'Towards an Embodied Understanding of the Structure/Agency Relationship'. *British Journal of Sociology*, 12, no. 1 (1999).

———. 'The Body and Difference'. In *Identity and Difference*, ed. Cathryn Woodward. London: Sage, 1997.

———. *The Body and Social Theory*. 1st edn. London: Sage Publications, 1993.

Siebers, Tobin. 'Disability Experience on Trial'. In *Material Feminisms*, ed. Stacey Alaimo and Susan Hekman. Bloomington: Indiana University Press. 2008.

Simmel, Georg. 'The Metropolis and Mental Life'. In *The Blackwell City Reader*, ed. Gary Bridge and Sophie Watson. Oxford: Wiley-Blackwell, 2002.

Smith, Dennis. 'The Civilizing Process and the History of Sexuality: Comparing Norbert Elias and Michel Foucault'. *Theory and Society*, 28, no. 1 (1999).

Tannen, Deborah. *Framing in Discourse*. Oxford: Oxford University Press, 1993.

Twigg, Julia. 'Carework as a Form of Bodywork'. *Ageing and Society*, 20, no. 4 (2000).

Wilchins, Riki. 'A Certain Kind of Freedom: Power and the Truth of Bodies—Four Essays on Gender'. In *GenderQueer: Voices from Beyond the Sexual Binary*, ed. Joan Nestle, Clare Howell, and Riki Wilchins. Los Angeles: Alyson Books, 2002.

Wright, Lawrence. *Clean and Decent: The Fascinating History of the Bathroom & the Water Closet, and of Sundry Habits, Fashions & Accessories of the Toilet, Principally in Great Britain, France, & America*. London: Routledge, 1963.

Young, Iris Marion. *On Female Body Experience: 'Throwing Like a Girl' and Other Essays*. Oxford: Oxford University Press, 2005.

Zerubavel, Eviatar. *The Elephant in the Room: Silence and Denial in Everyday Life*. Oxford: Oxford University Press, 2006.

INDEX

The Civilizing Process (Elias, Norbert), xiv, 74
Colebrook, Claire, 13–14, 36, 52
collective care in toileting: bodily openness, 141; connectedness, 141; empathy, lack of, 143; FASE (fear, anxiety, shame, and embarrassment), 140–142; social discomfort, 141–142; witnessing care, 142
conformity, 8, 14
connectedness, 140, 169–170
consciousness, 6, 16, 19, 46, 184
contact with toilet seat, 105
corpus infinitum: difference, 59–60; embodiment, 52; possibilities, 185; summary, 187
cruising in public toilets, 175–178

defecation, 73, 105
degendering, 53
Deleuze, Gilles, xix, 59
desire, social norm of, 36
de-territorialized, body, 34
difference, 59–60
Difference and Repetition (Deleuze, Gilles), xix
diffraction, xvi–xix, xx, xxiv, 46
disappearance, bodily, 16–19
discourse, 50
disembodiment: bodily borders, 15–16, 20, 62, 174; definition, 15; mind, training of, 16; movement and mental attention, 17; rational process, 16; selective attention, 17
dys-appearance, 17–19

economy of movement, 103, 107
Elias, Norbert: civilizing process, xiv; *The Civilizing Process*, xiv, 74; flaws in theory, 45; *homo clausus*, 3; hygiene and history, 85; posthumanizing, 61–62; social bonds, 133
embarrassment,. *See* FASE (fear, anxiety, shame, and embarrassment)
embodied ethics, 184
embodiment: abjection, 98; awareness, 20, 61, 117–118; body-selves, 54; daily life, 185–186; experience, 28; feminine, 115; phenomenology, 16;

sensory, 34, 134
The Emergence of Social Space: Rimbaud and the Paris Commune (Ross, Kristin), xxiii
emotions, xiii, 9–12, 19. *See also* FASE (fear, anxiety, shame, and embarrassment)
etiquette, 76–79
Europe, xiv
excretory shift, 77, 79, 84–88, 89n15

family makeup, 58
Farmer, Khadijah, 99
FASE (fear, anxiety, shame, and embarrassment): abjection, 98; body identity, 9–12; collective care in toileting, 142; creating, 167; and defecation, 105; excretion, 113; gender identity, 33; learned behavior, 12; naturalness, 12; overcoming, 164–166; power, lessening of, 62; public toilets, 120; relationships, lack of, 9; in women's public toilets, 105
fathers as caretakers, 144–146
fear. *See* FASE (fear, anxiety, shame, and embarrassment)
feelings on public toilet rules, xxiv
feminine sexuality, 7
feminism, xxvi, 49
feminist theory, 183–184
figurations, 28
flirting, 172–174
flush toilets, 72, 82
fragmentation, 29, 34–35

gay people. *See* queer/trans people
gender: cultural norms, 54; definition, xv; degendering, 53; identification, xiii; identity, xiii, 54–60, 133–137; natural, 32; nonconforming, 53–60, 99, 110; roles, 100; segregation, xiii, 80, 84, 98; social standards, xvi; toileting, 54; use of public toilets, 103–104
gender identity, 31–32
gender nonconforming : threat, masculinity, 110; policing in public toilets, 99, 113–114; women's public toilet, 111–112
gender segregation, 83

ABOUT THE AUTHOR

Dara Blumenthal is an interdisciplinary researcher, writer and creative thinker based in New York City. She completed her BA at the Gallatin School of Individualized Study at New York University, before moving onto doctoral research in sociology, followed by an MA in critical theory at the University of Kent, United Kingdom. While at Kent she held departmental scholarships in the School of Social Policy, Sociology and Social Research and the School of English.